MEDIA MANUALS

Panaflex Users' Manual

MEDIA MANUALS

Panaflex Users' Manual

David W. Samuelson

FOCAL PRESS
Boston   London

Focal Press is an imprint of Butterworth–Heinemann.

Library of Congress Cataloging-in-Publication Data

Samuelson, David W.
  Panaflex users' manual.
  (Media manuals)
  1. Panaflex motion picture camera. I. Title.
II. Series.
TR883.P36S26   1989   778.5′3   89-17133
ISBN 0-240-80052-4

British Library Cataloguing in Publication Data
Samuelson, David W.
  Panaflex users' manual.—(Media manuals).
  1. Cinematographic equipment : Cameras. Operation—
Manuals
I. Title II. Series
778.5′3

ISBN 0-240-80052-4

Butterworth–Heinemann
313 Washington Street
Newton, MA 02158–1626

10 9 8 7 6 5

Printed in the United States of America

Contents

THE DIRECTORS OF PHOTOGRAPHY'S PANAFLEX

THE CAMERA OPERATORS' PANAFLEX

THE CAMERA ASSISTANTS' PANAFLEX

USING PANAFLEX AND PANASTAR CAMERA ACCESSORIES

X

THE SOUND RECORDISTS' PANAFLEX

THE PRODUCTION MANAGERS' PANAFLEX

PANAVISION CREDITS

Foreword

When PANAVISION developed its "system" of cinematography, the goal was to create a system where crews could be absolutely assured that the cameras, lenses and all accessories would always fit together to provide a complete solution to their needs, irrespective of where in the world the equipment was being used. This consistency has enabled the equipment to become an integral part of the craft of filmmaking. Surprising as it may seem, up to this day in 1989, PANAVISION is the only system in the world where this applies.

From the very beginning, PANAVISION has not considered itself to be just another rental company, nor does it consider itself to be simply a manufacturing company. Our concept has always been to develop equipment in a partnership with the motion picture industry. We grew as a company by insisting that we be technically involved in the actual production of movies. This ongoing collaboration has allowed us to develop unique equipment that addresses the needs of the cinematographers and their crews. The majority of PANAVISION equipment has in fact been developed in response to the suggestions and ideas of filmmakers over the years, and we would like to thank these people for their contributions to the finely tuned systems described in the manual.

For PANAVISION's staff, this partnership with filmmakers has fostered a feeling of participation in the making of motion pictures, and therefore a sense of responsibility and pride in their own skills. The dedication felt by every employee at PANAVISION is far beyond the scope of this foreword, but it is the passion for the craft, and the continuous desire to provide the finest quality motion picture equipment and service, that has made PANAVISION what it is today.

Because PANAVISION doesn't sell the equipment it manufactures, the prime concern is to provide the highest quality and functionality, without the compromises required to meet a sales price. We can assure you that PANAVISION will continue its leadership for many years to come. It is our firm belief that the future of the motion picture industry is just as rich as its past. Our role, along with film manufacturers and laboratories, is to provide the cinematographer with an ever expanding palette of creative choices. The newest generation of PANAVISION Primo lenses is one illustration of the major advances we can all still achieve.

PANAVISION is honored to have David Samuelson write about our camera systems. The PANAFLEX USERS' MANUAL is a great accomplishment that could only be undertaken by someone with David's wealth of knowledge and experience. We would like to pay tribute to his willingness to give his time so that we all may be better informed and educated.

John Farrand
President and C.E.O.
PANAVISION Inc.
April 1989

Introduction

PANAVISION Inc. was founded by Robert E. Gottschalk in 1954, shortly after the introduction of the CinemaScope widescreen format, to fulfill the need for high-quality anamorphic projection lens attachments.

Within a year of the introduction of the CinemaScope format, ordinary 4 × 3 (1.37:1) pictures quickly looked old-fashioned, and theater owners frantically sought a source of good anamorphic lens attachments to enable them to show the new films without the need to modify their theaters or to be beholden to one supplier.

At that time Gottschalk owned a camera store in Westwood Village where he numbered among his customers many professional photographers and cinematographers. Among his acquaintances was an optical engineer who helped him to design a prism type de-anamorphoser which proved to be superior to the original CinemaScope projection lenses.

Within a short while he and a small staff, which included Frank Vogelsang, Tak Miyagishima, George Kraemer and Jack Barber, produced and delivered some 35,000 lenses until the market became saturated.

Other founder participants were Harry Eller, who owned the Radiant Screen Company in Chicago, the largest screen manufacturers in the U.S.; William Mann, an optical manufacturer; Richard Moore and Meridith Nicholson, both Directors of Photography; and Walter Wallin, an optical designer. All of this group dropped out of the company soon after the initial demand for projection lenses was satisfied.

In 1957, at about the same time that the demand for projection lenses was falling off, Gottschalk was asked by MGM to develop a set of anamorphic lenses with a 1.33:1 squeeze ratio for 65mm cameras for a forthcoming production, *Raintree County*, starring Elizabeth Taylor and Montgomery Clift, which attempted to outdo *Gone With the Wind*. The system was called CAMERA 65.

Another CAMERA 65 picture of the period was *Ben Hur* (1959), the first PANAVISION-lensed picture to win an Academy Award for Cinematography.

The CAMERA 65 system was later further developed, changed to a 1.25:1 squeeze ratio and called ULTRA PANAVISION.

Building upon this success, PANAVISION developed a system of non-anamorphic 65mm cameras and lenses, called SUPER PANAVISION. Such pictures as *Exodus, West Side Story, Lawrence of Arabia* and *My Fair Lady* all bore this label.

The next step was 35mm 2:1 anamorphic lenses and PANAVISION 35 was born. The lenses were called AUTO PANATARS, a name which has endured to this day.

These lenses incorporated patented counter-rotating focusing elements, developed by Walter Wallin, which eliminated what had become known as "anamorphic mumps," the swelling of faces in close-ups that upset many famous Hollywood actors and actresses and made them reluctant to appear in CinemaScope films.

Those who were in the screening room at MGM when the first tests of the new lenses were screened recalled that the entire audience clapped and cheered at what they saw. It was the dawning of a new age of anamorphic cinematography and since that time almost every truly major picture shot in the anamorphic format has been photographed using PANAVISION AUTO PANATAR lenses.

In 1958 PANAVISION Inc. received an Academy Award for Scientific and Technical Achievement for the development of these lenses.

Even Twentieth Century Fox, who pioneered the whole process of CinemaScope 2.35:1 anamorphic cinematography, quickly changed to PANAVISION lenses when they saw the improvement in image quality.

Among the early pictures shot using the PANAVISION 35 system and blown up to 70mm for road show presentation were *Beckett, The Cardinal* and *Doctor Zhivago.*

The next logical step was to provide cameras to go with their lenses and PANAVISION rapidly became well known for its innovative modifications to existing Mitchell and Arriflex cameras.

In the early 60's, with television making inroads into their traditional movie theater business and with so many of their pictures being photographed with cameras and lenses supplied by PANAVISION, MGM and many other major studios decided to close down their camera departments. Most sold off their entire inventory of cameras and lenses to PANAVISION.

This gave the company an abundance of Mitchell BNC 35mm cameras which they rebuilt as the PANAVISION SILENT REFLEX CAMERA, incorporating mirror shutter reflex viewfinding, crystal controlled motors, quietness of operation, lightness of weight and interchangeability of lenses among all cameras within the system, thus rendering all other Mitchell BNCs obsolescent. The PSR cameras were immensely successful and rapidly became the industry standard.

Robert Gottschalk realized that, good though the PSR was, it was not hand-holdable and set about designing and building a hand-holdable silent reflex support camera. The result was the PANAFLEX motion picture camera.

The rest, as they say, is history.

This book is dedicated to the memory of
Robert E. Gottschalk
1917–1982
and
Frank P. Vogelsang
1934–1988

Acknowledgements

My association with PANAVISION Inc. goes back to 1965. They were a small lens manufacturing and rental company in Los Angeles, and my brothers and I owned a small equipment rental company in London, and we became the first PANAVISION overseas representatives.

It was a turning point in all our lives. We have grown and matured together.

Since that time I have always had a close personal and working relationship with both the management and the technical staff at PANAVISION, and for this reason it was particularly gratifying, having retired from the Samuelson Group, to have been asked to become a consultant for the company.

Out of this special relationship has come this PANAFLEX USERS' MANUAL.

I wish to thank all at PANAVISION for their help, their co-operation and their knowledge, so freely shared, which has contributed to the writing of this, "their" book.

In particular I wish to thank John Farrand, the President and C.E.O. of Panavision, who gave me the opportunity, and Benjamin Bergery, who has been the Los Angeles coordinator whenever I have needed information from thousands of miles away.

Nearer home I wish to thank Karl Kelly of the Samuelson Group, who read an early version of the manuscript and made many useful suggestions.

Regrettably, it is no longer possible to thank Bob Gottschalk, who taught me to think the PANAVISION way. I only hope that he would have approved of what I have written.

David Samuelson
London
April 1989

Academy Technical or Scientific Awards won by PANAVISION Inc.
(1958–1978)

1958. Class II
"PANAVISION Inc. for the design and development of the Auto Panatar anamorphic photographic lens for 35mm Cinemascope photography."

1959. Class II
"DOUGLAS G. SHEARER of Metro-Goldwyn-Mayer Inc., and ROBERT E. GOTTSCHALK and JOHN R. MOORE of PANAVISION Inc. for the development of a system of producing and exhibiting wide-film motion pictures known as CAMERA 65."

1966. Class III
"PANAVISION Inc. for the design of the Panatron Power Inverter and its application to motion picture camera operation."

1967. Class III
"PANAVISION Inc. for a Variable Speed Motor for Motion Picture Cameras."

1968. Class II
"PANAVISION Inc. for the conception, design and introduction of a 65mm hand-held motion picture camera."

1969. Class III
"PANAVISION Inc. for the design and development of the Panaspeed Motion Picture Camera Motor."

1970. Class II
"PANAVISION Inc. for the development and engineering of the Panaflex motion picture camera."

1976. Class III
"PANAVISION Inc. for the design and development of super-speed lenses for motion picture photography."

1977. Class III
"PANAVISION Inc. for the concept and engineering of the improvements incorporated in the Panaflex Motion Picture Camera."

1977. Class III
"PANAVISION Inc. for the design of the Panalite, a camera-mounted controllable light for motion picture cameras."

1977. Class III
"PANAVISION Inc. for the engineering of the Panahead gearhead for motion picture cameras."

1978. Academy Award of Merit (Oscar)
"PANAVISION Inc. and its engineering staff under the direction of Robert E. Gottschalk, for the concept, design and continuous development of the Panaflex Motion Picture Camera System."

The
Film Producers'
PANAFLEX

PANAVISION Cameras for All Types of Film Making

The PANAVISION system of motion picture cinematography, and especially the PANAVISION PANAFLEX system, offers the Film Producer the widest possible range of creative possibilities and most efficient means of transferring a script to film. It matters not if that film is a multimillion dollar movie or a low budget project; a TV Series production, a TV Commercial or a Music Video; or whether it is shot in a local studio or on a faraway location. Whatever and wherever, to go PANAVISION is the most cost-efficient way to do it.

The PANAFLEX system
The PANAFLEX "SYSTEM" includes the following:

- The PLATINUM PANAFLEX, a hand-holdable, 35mm, truly silent, reflex camera.
- The very quiet GII GOLDEN PANAFLEX, the latest upgraded version of the original Academy Award winning PANAFLEX camera, incorporating many of the features of the PLATINUM PANAFLEX but at a greatly reduced price.
- The GOLDEN PANAFLEX, the workhorse camera of the PANAFLEX range.
- The PANAFLEX-X, an economical "second camera" which is similar to, and just as quiet as, a GOLDEN PANAFLEX but not hand-holdable.
- The PANASTAR, a state-of-the-art high speed camera for all purposes, including Special Effects.
- The PANAGLIDE floating camera and all manner of specialist items and systems, many of them unique to PANAVISION.

Other PANAVISION cameras
Other cameras available from PANAVISION include:

- The SUPER PSR (an advanced studio type camera based on the Mitchell NC), the regular PSR, now very inexpensive and much used for TV situation comedy multi-camera shoots.
- Various "special shot" cameras, including Mitchell S35 MK II's, Arriflex II's and III's and even B&H Eymos, all with hard PANAVISION lens mounts.
- Large format PANAVISION 65mm cameras for the ultimate in screen image quality.
- For 16mm filming PANAVISION has the studio-quiet PANAFLEX 16 camera.

All PANAVISION cameras have available for them a full range of compatible PANAVISION lenses.

2

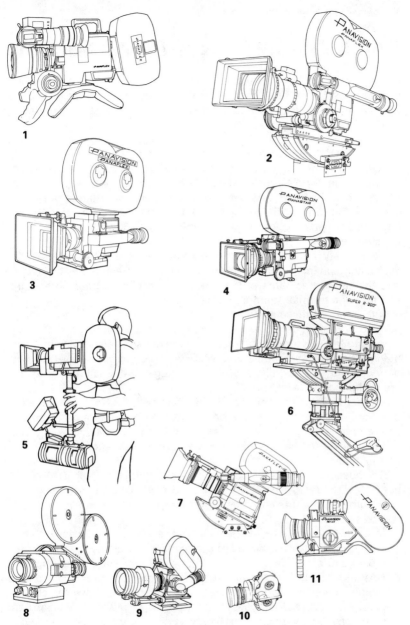

1. Hand-held PANAVISION PLATINUM PANAFLEX camera, 2. PANAVISION GII/ GOLDEN PANAFLEX camera, 3. PANAVISION PANAFLEX-X, 4. PANAVISION PANASTAR high-speed camera, 5. PANAVISION PANAGLIDE floating camera, 6. PANAVISION Super PSR studio camera, 7. PANAVISION PANAFLEX 16 camera, 8. PANAVISION Mitchell S35 camera, 9. PANAVISION Arri III camera, 10. PANAVISION Eymo camera, 11. PANAVISION 65mm hand-held camera.

The bigger the negative area the better.

PANAVISION Lenses for All Formats

Long before PANAVISION was known for its state-of-the-art cameras, it was famous for its fine lenses. Such films as *Lawrence of Arabia, Dr. Zhivago* and *My Fair Lady,* among others, won Oscars for Cinematography using PANAVISION crafted lenses.

PANAVISION's wide choice of formats and lenses

PANAVISION offers the greatest universe of presentation possibilities:

You can go "PANAVISION ANAMORPHIC" and have your picture photographed in the "2.35:1" anamorphic ratio giving your film a broad canvas. PANAVISION'S "E" series of AUTO PANATAR lenses are the very latest anamorphic lenses and the culmination of 30 years of continuous development.

Alternatively, for 2.35:1 release prints, you can go "PANAVISION SUPER-35," using spherical (non-anamorphic) lenses but extending the picture into the sound-track area of the negative to achieve greater image area (2/3 that of a normal Anamorphic frame).

With either of these systems you can have 70mm prints made for exhibition in large theatres and drive-in movie houses and when it comes to making transfers for cable, cassette, network and satellite TV, and even in-flight movie presentation, you will have a product where the close-ups of your major artists will be relatively as large on a small screen as they are in the theatre (see page 7).

Alternatively you can choose "PANAVISION SPHERICAL" using the regular Academy frame, which can be projected by any 35mm projector anywhere in the world, can be masked down to 1.66:1 or 1.85:1 and can be used for direct transfer to video for television presentation and for reduction to 16mm.

Whichever format you choose, the credits "FILMED IN PANAVISION" or "FILMED WITH PANAVISION CAMERAS AND LENSES" will say to your distributors, exhibitors and prospective filmgoers that yours is a carefully crafted movie.

PANAVISION's new PRIMO lenses are the first complete set of lenses especially conceived, designed and manufactured solely for motion picture usage to be commissioned in over a decade. The Primos highlight PANAVISION'S place at the forefront of lens manufacturers. These magnificent lenses have unique features about them which give the cinematographer the greatest possible scope for creativity, whether the photography be by natural or highly stylized lighting, or anything in between.

In conjunction with PANAVISION cameras, these state-of-the-art lenses provide the cinematographer with the ultimate image forming equipment.

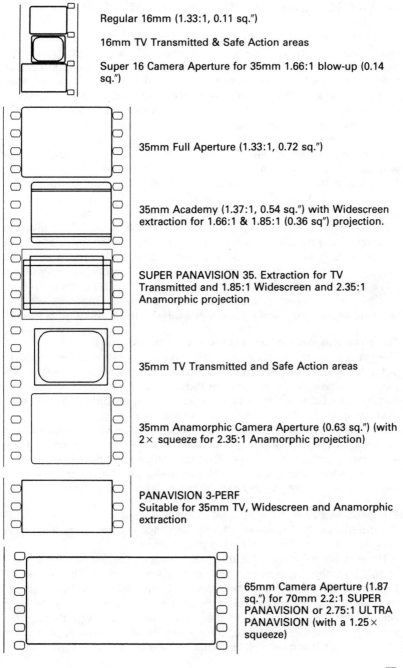

Regular 16mm (1.33:1, 0.11 sq.")

16mm TV Transmitted & Safe Action areas

Super 16 Camera Aperture for 35mm 1.66:1 blow-up (0.14 sq.")

35mm Full Aperture (1.33:1, 0.72 sq.")

35mm Academy (1.37:1, 0.54 sq.") with Widescreen extraction for 1.66:1 & 1.85:1 (0.36 sq") projection.

SUPER PANAVISION 35. Extraction for TV Transmitted and 1.85:1 Widescreen and 2.35:1 Anamorphic projection

35mm TV Transmitted and Safe Action areas

35mm Anamorphic Camera Aperture (0.63 sq.") (with 2× squeeze for 2.35:1 Anamorphic projection)

PANAVISION 3-PERF
Suitable for 35mm TV, Widescreen and Anamorphic extraction

65mm Camera Aperture (1.87 sq.") for 70mm 2.2:1 SUPER PANAVISION or 2.75:1 ULTRA PANAVISION (with a 1.25× squeeze)

5

A decision which is as old as CinemaScope itself.

Anamorphic versus Spherical

To shoot "anamorphic" or "spherical" is a technical decision the Producer must take. Put simply, "going anamorphic" involves photography with anamorphic lenses which squeeze a wide scene area onto a narrow film format. The film is subsequently projected in the theatre using a complementary anamorphic lens which unsqueezes the image to give a picture on the screen which is almost 2.4 times as wide as it is high. Television presentation is achieved by extracting a 1.33:1 image from the anamorphic original at the time the film is transferred to video tape.

Going "spherical" means photographing the film in a normal manner with an image which is a little more than 1⅓ times as wide as it is high (1.37:1) and from that losing a great deal of the top and bottom of the image for 1.85:1 widescreen presentation in the theatre. In this format almost all of the original image is used for television presentation.

A third alternative is to go "PANAVISION SUPER-35." This involves photographing the film with spherical lenses across the full width of the film, including the sound track, and from this using an optical printer to make an anamorphic internegative from which anamorphic theatrical release prints can be made. With the modern generation of fine grain film-stocks and improved optics, this is not an overwhelming problem. The tele-cine transfer to video can be made directly from the original format.

The pros and cons of anamorphic cinematography
The super-wide screen ratio of anamorphic presentation (called "2.35:1" but actually 2.4:1) is particularly good for pictures which contain many panoramic exterior scenes and for those which have many stretches of conversation between important characters, all of whom need to be seen in close-up to add importance to what they are saying.

An added advantage is that an anamorphic image can be "panned and scanned" at the time of tele-cine transfer to center the image on the most important element or character, thus maintaining the close-ups.

The disadvantage is the need to pan and scan. Purists argue that an electronic pan is not as aesthetically pleasant as a natural pan.

The pros and cons of spherical cinematography
The great advantage of spherical cinematography is that the same original image can be used for all forms of presentation.

The disadvantages are that the image must be framed top and bottom to suit both the widescreen and the television formats. This will entail additional cost in ceiling pieces, keeping lights higher and being careful that dolly tracks do not appear in the television picture. When shooting it will mean framing close-ups so that artists' heads do not go through the top of the widescreen format, thus losing much of their impact when shown on television with all the additional head and foot room.

6

The original scene

THE ANAMORPHIC ROUTE
V

THE SPHERICAL ROUTE
V

As seen in a Theatre
(2.35:1)

As seen in a Theatre
(1.85:1)

   < < > >   

As seen on television
(Panned and scanned from an
anamorphic original)

As seen on television
(Straight transfer from an
academy frame original)

7

PANAVISION's Special Equipment

PANAVISION is always interested in supplying custom engineered camera systems and accessories to suit the special needs of their customers. Over the years PANAVISION has designed and supplied many unique cameras, accessories and lenses from the special mirage effect lens that was used to shoot the desert mirage sequence in *Lawrence of Arabia* to various 35mm and 65mm 3D lenses and camera systems.

Cost-saving equipment
It has always been a part of PANAVISION'S philosophy to reduce costs and this has very often been achieved through innovation without sacrificing quality, usability or reliability.

Not least of the accessories which are available with all PANAVISION cameras are those which directly reduce production costs.

For productions where negative filmstock and processing costs are a significant part of the budget, PANAVISION can supply cameras fitted with their 3-PERF camera movement, saving 25% on these costs and reducing short-end wastage and camera re-loading time. The 3-PERF system, together with time code (see pages 216-227), is particularly suitable for film productions for TV presentation where the original negative is transferred directly to tape for post-production and distribution.

The PANAVID range of video-assist systems are well known and a great cost saver. For the PLATINUM PANAFLEX, PANAVISION engineers have developed an entirely new concept of dedicated flicker-free/freeze frame video assist system which uses a CCD image sensor of immense sensitivity and which scans in synchronization with the film, giving new possibilities for SFX shooting. The image quality of the CCD PANAVID is so high that it can be used for initial off-line video editing. The edited version can later be conformed with the film by the use of PANAVISION's AATONCODE time code system.

Another unique viewfinder accessory available from PANAVISION is the PANALUX, a military type night-sight device incorporated into an interchangeable PANAFLEX viewfinder. This enables the camera operator to see a scene clearly in even the darkest conditions, an enormous cost saver, especially when setting up large night exterior scenes, making it unnecessary to switch on the costly filming lights during the set-up and early rehearsal periods.

Laboratory equipment
Although relatively unknown to cinematographers, PANAVISION also supplies many leading film processing laboratories and special effects companies with a wide variety of optical printer lenses.

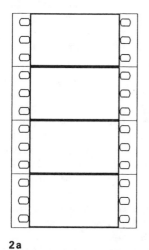

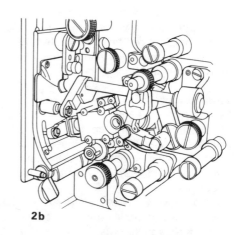

2a

2b

3

4

1. The special shimmer effect lens made by PANAVISION for the mirage sequence in *Lawrence of Arabia,* 2.a. PANAVISION 3-PERF film and b. camera movement, 3. PANAVISION PANAVID video viewfinder systems, 4. PANALUX night sight camera viewfinder system.

PANAVISION's New Technologies

Many of the new generation of PANAVISION's senior design personnel have had recent experience in the avionics, electronics and computer industries and have been successfully integrated into the design team to work alongside all those who are long established cinematographic engineers with many years of experience. With this formidable combination Film Producers can be assured that the latest technologies are available to them and that the means of making films has not been left to wallow in the horse and buggy era.

Time code

For Producers seeking to use new technology as a means to reduce post-production time and costs, PANAFLEX cameras can be fitted with PANAVISION AATONCODE time code system. This 35mm system is based upon the well established AATON 16mm system, which combines SMPTE time code with legible, "man-readable" information on every frame of negative and print.

If you have a PANAVISION AATONCODE time code generator fitted inside your camera, alongside each frame of film there will be a series of computer-readable dots, known as the "SMPTE time code." This "time stamp" records the precise date and time that the film was exposed (accurate to $1/24$ sec. over an 8 hour period), together with the Production No. (very useful when shooting an episodic TV production), the type of filmstock used and the camera No. and ID letter.

In addition, the date, time and "start frame," together with the Production No., the filmstock type and the camera No. and ID information, will be displayed in "man-readable" numbers and letters once every second.

The impact of the PANAVISION AATONCODE time-code system is greatest in the video post production stage of film making where the cost savings can be quite substantial.

PANAVISION welcomes the opportunity to discuss the new and emerging technology of time-code assisted post production with film and video editors, sound recordists and dubbing mixers, film laboratory managers, video transfer engineers, and all other film and video technicians who handle the image and sound, in whatever form, from when it leaves the camera until it reaches the screen as a finished product.

10

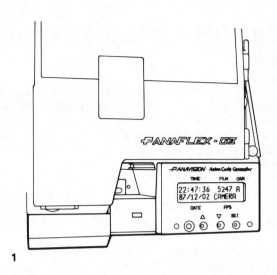

1

Direction of Travel →

        Date   Time   Start

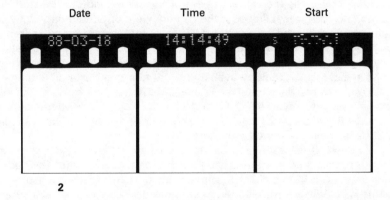

2

1. PANAVISION AATONCODE camera unit fitted to a PANAVISION GII camera,
2. PANAVISION AATONCODE time code film clip showing the man-readable
figures interspaced with the computer-readable digital time code information

PANAVISION's Production Safety Equipment

Since the beginning of Cinema the paying public has demanded their thrills in even more intense doses and the Motion Picture production industry has both stimulated and satisfied their needs. Fire, explosions, high-speed car chases and low-flying aircraft are the stuff that excites the audience and there is not a thrill experience invented that has not been filmed in PANAVISION.

This has led, in some cases, to film production becoming a hazardous occupation and, mindful of this, PANAVISION has produced special equipment that not only reduces the risks to all, on both sides of the camera, but also increases the creative possibilities to hype up the audiences.

Remotely controlled camera equipment

It is no coincidence that PANAVISION in Los Angeles, and many of its representatives worldwide, supply the Louma camera crane, the world's first remote control camera system, which married the technologies of the video-assist viewfinder and an electronically operated pan and tilt control system. This gives the camera operator the same field of view and the same operating capability as if the camera were on an ordinary dolly or crane with the crew all around it. The Louma camera crane makes it possible to put the camera into places and to have a flexibility of operation that would otherwise have been too hazardous or otherwise impossible.

To make this technology available to all film makers in all situations PANAVISION has developed the PANAREMOTE head control system. This system fits onto any PANAHEAD and allows the head to be operated from up to 150 feet away for reasons of safety or to get a shot which would not be possible by any other means.

The PANAHEAD is already the first choice among camera operators as the smoothest and most accurate pan and tilt head available. The added ability to use it from a distance with exactly the same feel, even using the same hand wheels, makes it an even more desirable item of equipment and expands the visual possibilities of film making without adding to the hazards.

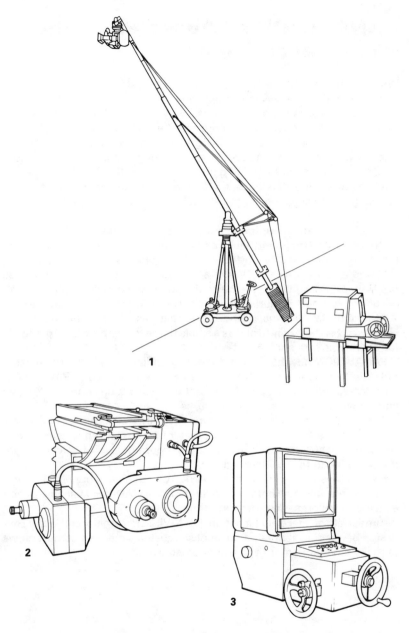

1. Louma camera crane with control console in the foreground, 2. Regular
PANAHEAD fitted with PANAREMOTE remote control motors, 3. PANAREMOTE
remote camera control console.

13

PANAVISION's "Added Production Value" Accessories

When comparing the costs of various camera systems the question most often asked is, "Does any extra cost show on the screen?" With regard to PANAVISION items the answer must always be "yes."

The fact that PANAVISION can provide the highest possible quality lenses, giving the possibility of much greater range of image control, and that PANAVISION cameras are quieter, making it possible to use more original sound, rather than doing looped voices at the dubbing stage, are obvious answers to this question. Others lie in the vast range of accessories that are available "off the shelf."

Typical accessories which are a part of the PANAVISION system
PANAVISION's equipment facilities are cram packed with special items that have been asked for in the past. Electronic control devices to synchronize a camera with a TV set, a computer, a video game, a process projector or HMI lighting, together with underwater housings, special lightweight or compact cameras all remain available for today's film makers.

Among the latest items are the special features of the PANAFRAME CCD video-assist camera, which allows a single frame to be frozen and "grabbed" for alignment with another set-up.

PANAVISION now has nodal camera mountings available that make it possible to mount any PANAFLEX camera nodally on any PANAHEAD. This accessory is particularly important when shooting certain SFX shots and for shooting cost saving miniatures.

Car mounts are available for all types of PANAFLEX cameras to enable cameras to be fitted to cars and other moving platforms, quickly and securely, so they can operate with a maximum degree of safety to both the camera and to those around it.

For the most spectacular shots PANATATE camera mounts, 360° turnover mounts which fit onto any regular PANAHEAD, are available to add a new dimension to any camera set-up.

Almost all of these advances in PANAVISION camera and accessory design have come about as a result of suggestions made by camera crews to make their work more efficient and creative.

14

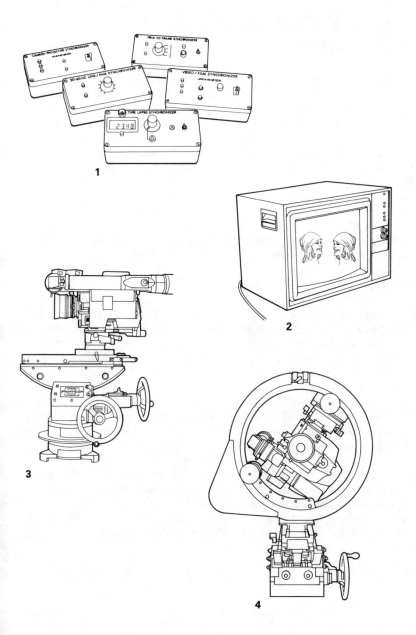

1. PANAFLEX electronic control boxes, 2. PANAFRAME video viewfinder system, 3. PANANODE nodal head attachment for the PANAHEAD, 4. PANATATE 360° turn-over mount.

15

PANAVISION and Future Technologies

PANAVISION cameras are continuously updated "state of the art" devices which incorporate more desirable features than any other motion picture camera. Every feature is there either because it makes pictures look better on the screen or because it makes it less expensive to make pictures, especially when shooting sync sound in confined conditions.

Development work never stops. The PANAVISION management team knows that if they were ever to sit back on their past laurels and not continue to spend vast sums of money on R&D they would quickly lose their leadership in the field. Because PANAVISION cameras and lenses are never sold but only leased on a picture-by-picture basis, PANAVISION is in the ideal position to constantly manufacture new items, to upgrade and service existing items and to withdraw those that become passé. You never get a noisy "yesterday's" camera from PANAVISION.

PANAVISION at the cutting edge of technology

For the ultimate in image quality, especially insofar as being able to have your picture projected using very much higher screen brightness levels you can shoot with PANAVISION cameras at 30 fps and because the cameras are so quiet in the first place you can do so without the camera noise becoming obtrusive. With this system images have freedom from flicker, faster pans and tilts are possible without strobing and there is finer grain and greater image detail.

Shooting at 30 fps is also advantageous for ultimate image quality when film is transferred to NTSC video; for shooting scenes which contain many in-shot video monitors and for transfer to High Definition TV (whichever systems may prevail in the future).

Irrespective of what generations of video origination, recording and display systems may be invented, and later relegated into disuse in the years to come, a transparent image on film will always be usable and marketable at any time in the future. This is more than can be said for productions that were recorded, even comparatively recently, on now obsolete video recorders. There is always a future for film.

16

1. PANAVISION's engineering design department, 2. PANAVISION's sound test room, 3. PANAVISION's environmental test chamber, 4. PANAVISION's MTF optical test bench, 5. Video formats may come and go but rolls of film last virtually forever.

17

Going PANAVISION puts a gloss on your production.

The Hidden Economies of "Going PANAVISION"

Every element of the PANAVISION system of Cinematography is inter-changeable. PANAVISION equipment is designed to minimize any pos-sibility of down time due to camera malfunction. PANAVISION's world-wide Distributor support network ensures that adequate back-up is never more than a phone call away.

Since PANAVISION cameras, lenses and related equipment are never sold and only leased on a picture-by-picture basis, the Producer can take only the minimum amount of equipment actually needed for a specific project and then only for the period it is required. For low-budget projects there are always lower cost options available and PANAVISION's client liaison personnel can always suggest alternative equipment whose bottom line cost will be less than any other.

Producers know that the very latest equipment, fully updated with all modifications, will be available to them and their production technicians; that equipment-for-equipment, PANAVISION has no equal; that the ser-vices offered have full back-up support; and that, in the final analysis, PANAVISION is the economic, reliable and responsible choice.

It is arguable as to whether it is more important to "go PANAVISION" on a big picture, when the financial stakes are greatest and visual impact on the screen must match the magnitude, or on a modest picture, when even a small holdup can cause expensive delays. PANAVISION has proven itself the most cost-effective camera system for films of all budgets.

"Going PANAVISION" is like taking out a low-cost insurance policy.

PANAVISION credits

More than 90% of the films at the top of *Variety*'s ALL-TIME FILM RENTAL CHAMPS list were photographed with PANAVISION lenses and cameras. With PANAVISION you are not guaranteed to have a box-office winner but it cannot hurt.

Since the first film to be filmed in PANAVISION, *Raintree County* in 1958, and the first PANAFLEX picture, Steven Speilberg's *Sugarland Express,* PANAVISION-equipped pictures have won more Oscars for Best Picture than any others using other lenses and cameras.

With PANAVISION you are always in good company.

18

Variety ALL-TIME FILM RENTAL CHAMPS
OF THE US-CANADA MARKET
Photographed with PANAVISION lenses and cameras (Jan 11th, 1989)

E.T. THE EXTRA-TERRESTRIAL	$228,618,939
STAR WARS	193,500,000
THE EMPIRE STRIKES BACK	141,600,000
GHOSTBUSTERS	130,211,324
JAWS	129,549,325
RAIDERS OF THE LOST ARK	115,598,000
INDIANA JONES AND THE TEMPLE OF DOOM	109,000,000
BEVERLY HILLS COP	108,000,000
BACK TO THE FUTURE	104,408,738
GREASE	96,300,000
TOOTSIE	95,296,736
THE EXORCIST	89,000,000
SUPERMAN	82,800,000
CLOSE ENCOUNTERS OF THE THIRD KIND	82,750,000
THREE MEN AND A BABY	81,313,000
BEVERLY HILLS COP II	80,857,776
GREMLINS	79,500,000
TOP GUN	79,400,000
RAMBO: FIRST BLOOD PART II	78,919,250
THE STING	78,212,000

"BEST PICTURE" Academy Award pictures
Photographed with PANAVISION lenses and cameras (1958-1988)

AMADEUS	Zaentz-Orion
ANNIE HALL	Rollins-Joffe-UA
THE APARTMENT	Mirisch-UA
BEN HUR	MGM
BUTCH CASSIDY AND THE SUNDANCE KID	20th Century Fox
THE DEER HUNTER	EMI-Cimino-Universal
DOCTOR ZHIVAGO	Ponti-MGM
THE FRENCH CONNECTION	D'Antoni-Schine-Moore-20th Century Fox
GANDHI	Indo-British Films-Columbia
THE GODFATHER PART II	Coppola Company-Paramount
THE GRADUATE	Embassy
IN THE HEAT OF THE NIGHT	Mirisch-UA
KRAMER vs. KRAMER	Columbia
LAWRENCE OF ARABIA	Horizon-Columbia
A MAN FOR ALL SEASONS	Highland-Columbia
M*A*S*H	20th Century Fox
MY FAIR LADY	Warner Bros
OLIVER!	Romulus-Columbia
ONE FLEW OVER THE CUCKOO'S NEST	Fantasy-UA
ORDINARY PEOPLE	Universal
PATTON	20th Century Fox
RAIN MAN	Gruber-Peters Company-UA
ROCKY	Chartoff-Winkler-UA
THE STING	Bill/Phillips-Hill-Zanuck/Brown-Universal
TERMS OF ENDEARMENT	Brooks-Paramount
WEST SIDE STORY	Mirisch-UA

Academy Accolades

Since 1960, when the first films shot with PANAVISION lenses, *Anatomy of a Murder, Ben Hur* and *The Diary of Anne Frank,* received Academy Award Nominations for Best Picture of the Year, no less than 83 out of a possible 145 such pictures have been so honored, 24 of them winning.

**Producers who have won Oscars and Oscar Nominations for films photographed with PANAVISION cameras and/or lenses:
(winners in bold)**

IRWIN ALLEN	The Towering Inferno
ROBERT ALTMAN	Nashville
TAMARA ASSAYEV	Norma Rae
RICHARD ATTENBOROUGH	**Gandhi**
ROBERT ALAN AURTHUR	All That Jazz
WARREN BEATTY	Heaven Can Wait
CLAUDE BERRI	Tess
TONY BILL	**The Sting**
WILLIAM PETER BLATTY	The Exorcist
JOHN BOORMAN	Deliverance
JOHN BRABOURNE	A Passage to India
JAMES L. BROOKS	**Terms of Endearment**
DAVID BROWN	The Verdict
DAVID BROWN	Jaws
TIMOTHY BURRILL	Tess
ROBERT CHARTOFF	The Right Stuff
ROBERT CHARTOFF	**Rocky**
MICHAEL CIMINO	**The Deer Hunter**
WALTER COBLENZ	All the President's Men
ROBERT F. COLESBERRY	Mississippi Burning
FRANCIS FORD COPPOLA	American Graffiti
FRANCIS FORD COPPOLA	**The Godfather Part II**
PHILIP D'ANTONI	**The French Connection**
MICHAEL DEELEY	**The Deer Hunter**
MICHAEL DOUGLAS	**One Flew Over the Cuckoo's Nest**
ROBERT EVANS	Chinatown
EDWARD S. FELDMAN	Witness
JOHN FOREMAN	Prizzi's Honor
JOHN FOREMAN	Butch Cassidy and the Sundance Kid
CARL FOREMAN	The Guns of Navarone
MELVIN FRANK	A Touch of Class

Producers who have won Oscars and Oscar Nominations for films photographed with PANAVISION cameras and/or lenses (continued):

GRAY FREDERICKSON	**The Godfather Part II**
STEPHEN J. FRIEDMAN	The Last Picture Show
BRUCE GILBERT	On Golden Pond
RICHARD GOODWIN	A Passage to India
HOWARD GOTTFREID	Network
RICHARD GREENHUT	Hannah and Her Sisters
MICHAEL GRILLO	The Accidental Tourist
JEROME HELLMAN	Coming Home
NORMA HEYMAN	Dangerous Liaisons
ROSS HUNTER	Airport
STANLEY R. JAFFE	Fatal Attraction
STANLEY R. JAFFE	**Kramer vs. Kramer**
NORMAN JEWISON	A Soldier's Story
NORMAN JEWISON	Fiddler on the Roof
NORMAN JEWISON	The Russians Are Coming
CHARLES H. JOFFEE	**Annie Hall**
MARK JOHNSON	**Rain Man**
QUINCY JONES	The Color Purple
LAWRENCE KASDAN	The Accidental Tourist
KATHLEEN KENNEDY	The Color Purple
KATHLEEN KENNEDY	E.T. The Extra-Terrestrial
STANLEY KRAMER	Guess Who's Coming to Dinner?
STANLEY KRAMER	Ship of Fools
GARY KURTZ	Star Wars
GARY KURTZ	American Graffiti
SHERRY LANSING	Fatal Attraction
ARTHUR LAURENTS	The Turning Point
JOSHUA LOGAN	Fanny
FRANK MARSHALL	The Color Purple
FRANK MARSHALL	Raiders of the Lost Ark
FRANK McCARTHY	**Patton**
HOWARD G. MINSKY	Love Story
WALTER MIRISCH	**In the Heat of the Night**
HANK MOONJEAN	Dangerous Liaisons

Producers who have won Oscars and Oscar Nominations for films photographed with PANAVISION cameras and/or lenses
(continued):

RALPH NELSON	Lilies of the Field
CHARLES OKUN	The Accidental Tourist
PATRICK PALMER	Children of a Lesser God
PATRICK PALMER	A Soldier's Story
JOHN PEVERALL	**The Deer Hunter**
JULIA PHILLIPS	**The Sting**
MICHAEL PHILLIPS	**The Sting**
MARTIN POLL	The Lion in Winter
SIDNEY POLLACK	Tootsie
CARLO PONTI	Dr Zhivago
INGO PREMINGER	M*A*S*H*
OTTO PREMINGER	Anatomy of a Murder
ROBERT B. RADNITZ	Sounder
BOB RAFELSON	Five Easy Pieces
DICK RICHARDS	Tootsie
FRED ROOS	**The Godfather Part II**
ALEX ROSE	Norma Rae
AARON ROSENBERG	Mutiny on the Bounty
HERBERT ROSS	The Turning Point
ROBERT ROSSEN	The Hustler
RICHARD ROTH	Julia
ALBERT S. RUDDY	**The Godfather**
JONATHAN SANGER	The Elephant Man
BERNARD SCHWARTZ	Coal Miner's Daughter
RONALD L. SCHWARY	A Soldier's Story
RONALD L. SCHWARY	**Ordinary People**
MICHAEL SHAMBERG	The Big Chill

22

Producers who have won Oscars and Oscar Nominations for films photographed with PANAVISION cameras and/or lenses
(continued):

STEVEN SPEILBERG	The Color Purple
STEVEN SPEILBERG	E.T. The Extra-Terrestrial
SAM SPIEGEL	Nicholas and Alexandra
SAM SPIEGEL	**Lawrence of Arabia**
RAY STARK	The Goodbye Girl
RAY STARK	Funny Girl
GEORGE STEVENS	The Diary of Anne Frank
BURT SUGARMAN	Children of a Lesser God
LAWRENCE TURMAN	The Graduate
HAL B. WALLIS	Anne of a Thousand Days
HAL B. WALLIS	Beckett
JACK L. WARNER	**My Fair Lady**
RICHARD WECHSLER	Five Easy Pieces
BILLY WILDER	**The Apartment**
IRWIN WINKLER	The Right Stuff
IRWIN WINKLER	**Rocky**
ROBERT WISE	The Sand Pebbles
ROBERT WISE	**West Side Story**
JOHN WOOLF	**Oliver!**
SAUL ZAENTZ	**Amadeus**
SAUL ZAENTZ	**One Flew Over the Cuckoo's Nest**
DARRYL F. ZANUCK	The Longest Day
RICHARD D. ZANUCK	The Verdict
RICHARD D. ZANUCK	Jaws
SAM ZIMBALIST	**Ben Hur**
FRED ZINNEMANN	**A Man for All Seasons**
FREDERICK ZOLLO	Mississippi Burning
	. . . and more to come

The choice is wide but it can be narrowed down.

Selecting PANAVISION Equipment

When you decide to shoot a film with PANAVISION equipment you have the option of many alternative types of equipment with an equally wide range of prices appropriate to all types of productions and budgets. PANA-VISION'S customer liaison personnel are always available and ready to guide you through all the choices, to discuss the costs with you and to quote for you the most suitable and best equipment your budget can support. Nothing is impossible and never assume that PANAVISION equipment is too expensive for your picture.

Conditions of business

PANAVISION equipment is leased subject to standard production-by-production contracts, copies of which are available upon request.

The attention of all clients is drawn to the fact that PANAVISION Inc. and its Distributors supply equipment, materials and services only in accordance with their respective conditions of business. Copies of these Conditions of Business, *which include Clauses which exclude, limit or modify the liability of the company and provide for an indemnity from the customer in certain circumstances* are available on request.

Typical PANAVISION Equipment Choices

At first glance the variety of choices of PANAVISION equipment options is overwhelming; but by the time the pros and cons of the various demands and aspirations of the script have been weighed by the Director and Director of Photography, the most economical option begins to come clear. To help you in your choice the PANAVISION staff are always ready to discuss your requirements and to supply a well documented catalog.

Here is a headline guide to some of the possible choices:

The very finest, and quietest, camera system in the world:
PANAVISION PLATINUM PANAFLEX camera

The latest version of the long established camera system:
PANAVISION GII/GOLDEN PANAFLEX camera

The non-handholdable PANAFLEX system:
PANAVISION PANAFLEX-X camera

Low cost studio style cameras for low-budget pictures:
SUPER PANAVISION SILENT REFLEX camera

High speed cameras:
PANAVISION PANASTAR 2-120fps camera

PANAVISION PRIMO lenses, the best there are:
10, 14.5, 17.5, 21, 27, 35, 40, 50, 75, 100, 150 & 200mm T1.9
17.5-75mm T2.3 zoom lenses

A selection of PANAVISION's enormous range of lenses:
14-180mm "Z" series lenses (Zeiss glass, PANAVISION mechanics)
14-150mm PANAVISION ULTRA SPEED T1.0-T1.9 lenses
8-1000mm PANAVISION NORMAL SPEED LENSES
20-100mm T3.1 COOKE/PANAVISION 5:1 ZOOM lens
25-250mm T4 COOKE/PANAVISION 10:1 SUPER PANAZOOM
24-2000mm PANAVISION ANAMORPHIC LENSES

Desirable additional equipment:
PANAVISION PANAHEAD geared tripod head
Sealed lead acid batteries with built-in chargers
Reversible magazines for the PLATINUM and PANASTAR PANAFLEXES
PANAVISION PANAGLIDE floating camera system
PANAVISION CCD Flicker free PANAVID video system with frame store
PANAVID video systems
PANAVISION/AATONCODE TIME CODE system

The
Film Directors'
PANAFLEX

The "Go PANAVISION" Decision

To "go PANAVISION" is an early and an important decision on any film.

With the security of PANAVISION the Director is assured that, so far as the camera is concerned, he has the best possible chance of delivering the film on time, within budget and with the maximum image impact on the screen.

In the environment of a film production set the omnipresence of a friendly camera can contribute much to creating an atmosphere of quiet professionalism. The PANAFLEX camera is sleek and small and doesn't get in the way. It has been a design aim of the new PLATINUM PANAFLEX that it should be "transparent" on the set.

PANAVISION cameras don't need blankets over them to keep the Sound Recordist quiet

Always quiet and efficient, PANAFLEX cameras do not need blankets thrown over them to pacify the sound recordist.

Changing a lens on a PANAFLEX is simple and swift; nothing extra is needed to bottle up the camera noise, so your camera crew will not mind how many alternative setups you try before you settle on just the right one.

Somehow, there is a touch of humanity about the PANAFLEX camera which helps to keep your actors and actresses relaxed. Maybe it is the soft contours or its discrete color, maybe because it is only seen and never heard, or maybe it's because the crew never has to take it apart just when you are ready to go for an important take. PANAFLEX cameras never draw attention to themselves.

For whatever reason, everyone gets on well together around a PANAFLEX.

Customized equipment

PANAVISION has a long history of supplying hardware solutions to shooting problems. Typical examples are:

A lightweight AUTO PANATAR anamorphic lens to use on a skydiver's helmet.

A rifle sight that can be deployed in front of the camera lens during a take.

A mesmerizer lens to create a mystical dreamlike experience.

A camera body on a self-propelled platform to run under a big truck.

A super-lightweight camera to use with Garrett Brown's Skycam system.

Various single and two-camera 3D systems.

If your script has some special equipment need to make a point more effectively talk to PANAVISION. The chances are they will be able to come up with a solution. . . .

28

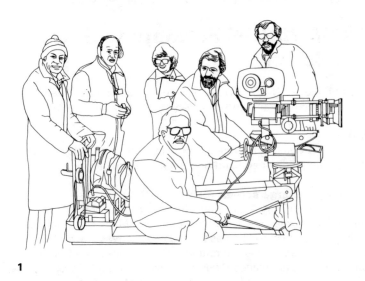

1

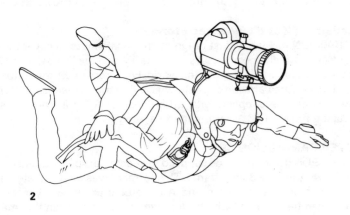

2

1. A PANAFLEX camera is a central part of the film making team, 2. A PANAVISION AUTO PANATAR anamorphic lens mounted on a lightweight camera on a skydivers' helmet.

The most important ingredient of a successful film is the script.

PANAVISION Formats to Suit the Script

The range of PANAVISION camera equipment is vast. There is no script or budget camera demand that PANAVISION cannot satisfy.

65mm
PANAVISION has a full range of 65mm cameras with reflex, hand-holdable, studio-quiet, high-speed and underwater capabilities.

Anamorphic large screen formats
35mm PANAVISION Anamorphic lenses are readily available for pictures destined for large screen theatrical release. Many Directors like to shoot in this format because it gives them a wide canvas that makes their movie more special in the theatre and yet gives a sense of intimacy to a group shot. Anamorphic scenes retain the power and the impact of close close-ups by using "pan and scan" techniques at the telecine stage when a 1.33:1 image is extracted for the small screen. The decision to "go AN-AMORPHIC" or even "go 65mm" is no problem with PANAVISION.

Academy, TV and Widescreen formats
PANAVISION has the widest range of spherical lenses to cover the Academy frame, which encompasses both the TV and the Widescreen (1.85:1) formats, available anywhere in the world. For Directors who wish to ensure that their pictures are correctly shown on theatrical screens PANAVISION cameras may be supplied with a hard matte. A TV image can still be extracted from a 1.66:1 hard-matted negative.

SUPER PANAVISION 35
PANAVISION cameras can be supplied to shoot with spherical lenses across the full width of the 35mm frame. This format gives an adequate negative area for large screen 70mm and Anamorphic prints and yet the video format can be extracted without the need to pan and scan the image.

PANAVISION 3-PERF
PANAVISION's 3-PERF three perforation pull-down system reduces film production costs by saving more than 25% of all camera filmstock costs and processing charges. It is especially suitable for films destined only for video post-production and presentation.

Regular 16 and Super 16
For both Regular 16 and Super 16 shooting, where the sound track is included in the 16mm frame area to give a larger negative for 1.66:1 and 1.85:1 35mm blow-up prints, the PANAFLEX 16 is the ideal camera.

30

1. 65mm

2. Typical scene relative to a 2.35:1 frame.

3. The same scene relative to a TV Transmitted area frame.

4. Head & shoulders C.U. relative to a 1.85:1 frame.

5. Same size C.U. relative to a TV Transmitted Area frame.

6. 1.85:1 image area relative to an Academy frame.

7. TV Transmitted area relative to an Academy frame.

8. 1.85:1 and TV Transmitted relative to an Academy frame.

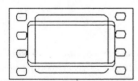

9. 2.35:1, 1.85:1 & TV image areas relative to a SUPER PANAVISION 35 frame.

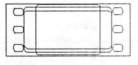

10. 2.35:1, 1.85:1 & modified TV image areas relative to a PANAVISION 3-PERF frame.

11. 1.66:1 Super 16 image area.

12. 1.66:1 image area derived from Regular 16.

13. Comparison of 1.66:1 image areas derived from Super 16 and Regular 16.

PANAVISION Cameras to Suit the Script

PANAVISION is able to supply the quietest hand-holdable sync camera, floating cameras, high-speed or single-shot cameras, plate cameras, underwater cameras, and so on. PANAVISION has them all, off the shelf.

PANAFLEX cameras

PANAVISION's computer-age camera is the PLATINUM PANAFLEX, an advanced technology instrument to make filming easier, quicker and quieter.

The GII GOLDEN PANAFLEX is an updated version of the GOLDEN at an economy price. It is very quiet and can be used with all the same lenses and most of the same accessories as the PLATINUM.

The GOLDEN PANAFLEX is the standard camera of the PANAFLEX range of cameras. It has practically all of the features required for general day-to-day cinematography, is quiet enough for most shooting environments and is very competitively priced compared to any other make of camera.

In close support of these cameras is the PANAFLEX-X, basically similar to the GII/GOLDEN PANAFLEXES but with a fixed viewfinder. It does, however, use the same magazines, matte boxes and most other accessories and is proportionately less expensive. Directors often choose an X as a second camera.

Complementing the PANAFLEX range of cameras are the PLATINUM PANASTAR and the PANASTAR single frame to 120 fps cameras especially designed for action unit and special effects cinematography. Where extraordinary image steadiness, the choice of ultra fast and ultra slow camera speeds and camera ruggedness are all at a premium, the "STARS" perform perfectly. The PLATINUM STAR is incredibly quiet for an SFX camera and at 24 fps is quite suitable for second camera use in many exterior sync-sound shooting conditions.

SUPER PSR cameras

Even less expensive than the X is the Super PSR camera. Based on Mitchell movements these cameras are the answer when the overriding need is for a studio camera at the least possible cost. They are particularly useful for multi-camera shoots including TV Situation Comedies.

PANAVISED support cameras

PANAVISION can supply a wide variety of Arriflex and Mitchell cameras, all fitted with PANAVISION lens mounts to make all cameras take all PANAVISION lenses.

1. The PLATINUM PANAFLEX
The quietest camera in the world
and incorporating every possible
"Production Value" facility

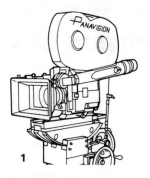

1

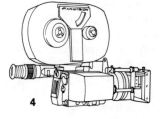

2

2. The GII/GOLDEN PANAFLEX
A much quieter version
of an old friend

3. The PANAFLEX-X
An economy version of the PANAFLEX

3

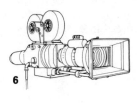

4

4. The PANASTAR
6 to 120 fps
Crystal controlled 12 to 120 fps

5. The Super PSR
Especially good for multi-camera shooting

5

6. "Pan" Mitchells & Arris
Ideal as support cameras

6

PANAVISION Lenses to Suit the Script

The scene-by-scene choice of lenses can make an enormous difference to the way a story is told. PANAVISION has lenses in profusion.

PANAVISION spherical lenses

Top of the range of non-anamorphic lenses is the fantastic new PRIMO series of lenses. Their undoubted superiority in definition, contrast, freedom from distortion and color balance between lenses makes them the No. 1 choice not only when the script demands crisp, gutsy images but also when the requirement is to produce soft and subtle images in conditions of harsh backlighting, diffusion and smoke.

Of particular interest to the Director is the variety of very fast lenses. Although it is not quite correct to say that, in combination with the new fast filmstocks, they see in the dark, they almost do. For the Director this opens up tremendous creative possibilities. PANAVISION has several lenses that are T1.0, T1.1 or T1.2. All unique. The creative possibilities of low-key lighting can be very evocative and PANAVISION wide-aperture lenses give you the opportunity to add this to your ability to tell a story. They have a further advantage in that there are many occasions when scenes can be shot which would otherwise be impossible for lack of lighting or when lighting time can be reduced or even saved.

PANAVISION also has some very special wide-angle lenses, including the fantastic PRIMO distortion-free 10mm T1.9.

PANAVISION zoom lenses use proprietary glass elements, mostly Cooke, but have PANAVISION designed and engineered mechanics to ensure the smoothest zoom movements with no juddering or image jumping effects when you zoom in and out.

The PRIMO 17.5-75mm T2.3 zoom lens is the flagship of the PANAVISION zoom lens range providing prime lens performance characteristics with zoom lens flexibility.

PANAVISION anamorphic lenses

For films designed to be blown-up to 70mm for very large screen presentations the camera optics need to be good in the first place. The fact that PANAVISION anamorphic lenses have been used for almost every major epic, as well as for the majority of low-budget anamorphic films, during the past 30 years, and still continue to be so, is proof in itself.

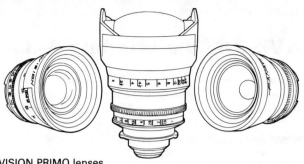

1. PANAVISION PRIMO lenses

FOCAL LENGTH	LENS ANGLES (Academy camera aperture)	
	HORIZONTAL	VERTICAL
200mm	6.2°	4.5°
150mm	8.3°	6.1°
100mm	12.5°	9.1°
75mm	16.6°	12.1°
50mm	24.7°	18.1°
40mm	30.6°	22.6°
35mm	34.8°	25.7°
27mm	44.2°	33.0°
21mm	55.1°	41.7°
17.5mm	64.1°	49.1°
14.5mm	74.2°	57.7°
10mm	95.3°	77.3°

2. PANAVISION PRIMO lens focal length/lens angle chart

3. PANAVISION AUTO PANATAR anamorphic lens

35

What you see through the viewfinder should be what you get on the screen.

PANAVISION Viewfinding Aids

The part of the film camera with which the Director has the closest relationship is the viewfinder. It therefore matters to him that the viewfinder system, and the video assist system, if fitted, are bright and clear.

Optical viewfinder systems

The optical viewfinders on all PANAVISION PANAFLEX cameras are particularly bright and are sometimes referred to as being "brighter than life."

All PANAFLEX viewfinders incorporate a PANAGLOW illuminated ground-glass reticle which enables the markings to be seen, even in the dark.

All PANAFLEX and PSR cameras (except the "X") have a viewfinder magnification system which enables the Director to zoom in on the ground glass image to take a closer look at a particular part of the scene.

All PANAFLEX and PSR viewfinders have a white bezel around the eye-piece focus ring where individuals can mark their own personal focus setting.

A PANALUX eyepiece, which incorporates a military night-sight device, may be used in place of the regular PANAFLEX eyepiece as an aid to setting up night scenes. It is particularly useful for large night exterior scenes when all the crowd arranging can be done without the filming lights being switched on.

Another useful accessory is the PANAFINDER portable viewfinder which can be used with any lens and which the Director can carry around and hold in his hand as an aid to determining a camera setup.

PANAVID video assist viewfinders

PANAVID video-assist systems can be fitted to any PANAFLEX camera.

The PANAVID CCD system gives high-quality, flicker-free images, even in very low light levels. It incorporates a FRAME GRAB facility to exactly superimpose the present setup with a previous one. The quality of the CCD system is good enough to use for off-line video editing.

All PANAVID video assists can be made to produce flicker-free pictures by exchanging the camera's mirror shutter for a fixed-pellicle mirror, a feature unique to PANAVISION.

36

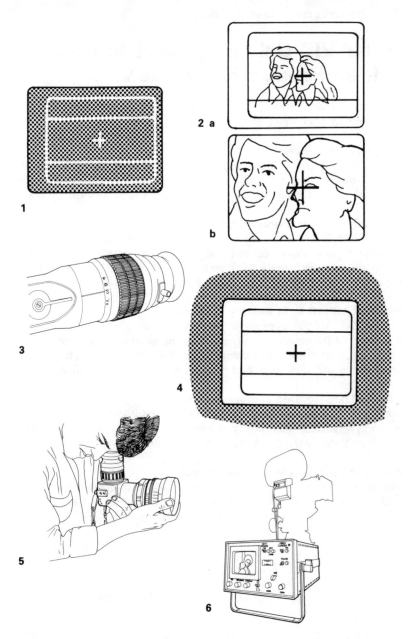

1. PANAGLOW illuminated ground glass markings, 2.a. Normal image though the viewfinder showing the ground glass markings, 2.b. Enlarged image through the viewfinder, 3. PANAFLEX eyepieces have a place to make personal setting marks, 4. PANALUX viewfinders can see in the dark, 5. PANAFINDER viewfinder, 6. PANAVID video assist system.

37

PANAVISION Equipment to Give Added Production Values

PANAVISION not only supplies cameras and lenses but can also supply a wide range of complementary specialist equipment.

Image control filters
PANAVISION has developed some very special matte boxes beyond the normal multi-stage filter holders. To give added production values PANAVISION can make filters slide, rotate and tilt, all in-shot, if required.

Inclining prism low-angle attachment
An INCLINING PRISM placed in front of the taking lens enables a shot to be taken from ground level as though a hole had been cut in the floor to accommodate the camera. With an INCLINING PRISM there is no loss of exposure, definition or image orientation.

PANAFLASHER light overlay accessory
The PANAFLASHER gives the film a very low overall exposure, which has the effect of increasing the exposure in the shadow areas and reducing image contrast. If the light is colored it has the effect of tinting the shadows and leaving the highlights, and most of the skin tones, unaffected.

PANAGLIDE floating camera systems
For gravity-free camera movements a PANAFLEX PANAGLIDE is the choice when sync sound recording is important, as is a PANARRI PANAGLIDE for when transportability is the primary consideration.

PANATATE turnover camera mount
The interpretation of a script sometimes calls for creative camera angles and even the camera turning over in space. This can be achieved with the PANATATE turnover mount, which fits onto any PANAHEAD.

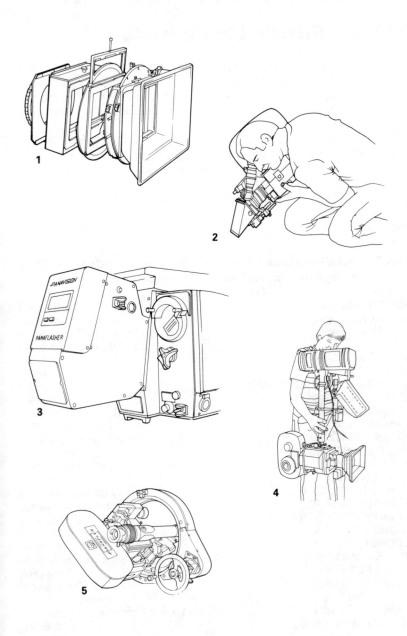

1. PANAVISION special facilities 6.6″ matte box, 2. INCLINING PRISM low-angle lens attachment, 3. PANAFLASHER light overlay accessory, 4. PANAGLIDE floating camera systems, 5. PANATATE turnover camera mount.

PANAVISION's Credentials

The Director's job is all about concentration. Concentration on the artists, concentration on interpreting the script and how a sequence is finally going to cut together on the screen, concentration on maintaining the mood to keep the audience absorbed, and so on. He or she has a great deal to think about.

For this reason it is of the utmost importance to the Director that all involved in the making of the picture, including the camera and sound crews, are efficient and equipped to give their best. The results of this will undoubtedly show on the screen.

From David Lean to Steven Spielberg there is not a great contemporary Director of motion picture films who has not put his trust in PANAFLEX cameras and PANAVISION lenses. Directors can have confidence with PANAVISION.

Directors who have won Oscars and Oscar Nominations for films photographed with PANAVISION cameras and/or lenses
(Winners in **bold**)

WOODY ALLEN	**Annie Hall**
WOODY ALLEN	Interiors
WOODY ALLEN	Broadway Danny Rose
WOODY ALLEN	Hannah and Her Sisters
ROBERT ALTMAN	M*A*S*H*
HAL ASHBY	Coming Home
RICHARD ATTENBOROUGH	**Gandhi**
JOHN G. AVILDSEN	**Rocky**
HECTOR BABENCO	Kiss of the Spider Woman
WARREN BEATTY	Heaven Can Wait
ROBERT BENTON	**Kramer vs. Kramer**
BRUCE BERESFORD	Tender Mercies
PETER BOGDANOVITCH	The Last Picture Show
JOHN BOORMAN	Deliverance
JAMES L. BROOKS	**Terms of Endearment**
MICHAEL CIMINO	**The Deer Hunter**
CHARLES CRICHTON	A Fish Called Wanda
GEORGE CUKOR	**My Fair Lady**
FEDERICO FELLINI	Satyricon
MILOS FORMAN	**One Flew Over the Cuckoo's Nest**
MILOS FORMAN	**Amadeus**
WILLIAM FRIEDKIN	The Exorcist
BUCK HENRY	Heaven Can Wait
GEORGE ROY HILL	Butch Cassidy and the Sundance Kid
GEORGE ROY HILL	**The Sting**
JOHN HOUSTON	Prizzi's Honor
NORMAN JEWISON	Fiddler on the Roof

40

Directors who have won Oscars and Oscar Nominations for films photographed with PANAVISION cameras and/or lenses,
continued:

STANLEY KUBRICK	2001: A Space Odyssey
AKIRA KUROSAWA	Ran
DAVID LEAN	**Lawrence of Arabia**
DAVID LEAN	Doctor Zhivago
DAVID LEAN	A Passage to India
BARRY LEVINSON	**Rain Man**
GEORGE LUCAS	Star Wars
SIDNEY LUMET	Dog Day Afternoon
SIDNEY LUMET	Network
SIDNEY LUMET	The Verdict
ADRIAN LYNE	Fatal Attraction
JOSEPH L. MANKIEWICZ	Sleuth
MIKE NICHOLS	**The Graduate**
MIKE NICHOLS	Silkwood
ALAN J. PAKULA	All the President's Men
ALAN PARKER	Midnight Express
ALAN PARKER	Mississippi Burning
ROMAN POLANSKI	Chinatown
SYDNEY POLLACK	They Shoot Horses, Don't They?
SYDNEY POLLACK	Tootsie
ROBERT REDFORD	**Ordinary People**
CAROL REED	**Oliver**
MARTIN RITT	Hud
JEROME ROBBINS	**West Side Story**
HERBERT ROSS	The Turning Point
MARK RYDELL	On Golden Pond
FRANKLIN J. SCHAFFNER	**Patton**
JOHN SCHLESINGER	Darling
JOHN SCHLESINGER	Sunday Bloody Sunday
MARTIN SCORSESE	Raging Bull
STEVEN SPIELBERG	Close Encounters of the Third Kind
STEVEN SPIELBERG	Raiders of the Lost Ark
STEVEN SPIELBERG	E.T. the Extra-Terrestrial
FRANCOIS TRUFFAUT	Day for Night
PETER WEIR	Witness
ROBERT WISE	**West Side Story**
WILLIAM WYLER	**Ben Hur**
FRED ZINNEMANN	Julia
	. . . and more to come

41

The
Directors of Photography's
PANAFLEX

The Directors of Photography's PANAFLEX

Although the Producer and the Director may feel especially comfortable selecting PANAVISION cameras and lenses, the decision to "go PANAVISION" is crucial to the Director of Photography.

PANAVISION has the widest range of lenses available anywhere. The widest apertures, the widest angles, the longest telephotos, the broadest choice of zooms. Lenses that have good contrast and excellent resolution and lenses that are as free of distortion, color aberration and flare as it is possible to be. Good lenses and lots of them.

PANAFLEX cameras have many wonderful features. They are exceptionally quiet, even for close close-ups; they can be tripod or dolly mounted or hand held; they can be compact or low profile and have three sizes of magazines, any of which can be used either on top of the camera body or at the rear. It's not just a camera; it's a comprehensive system for cinematography.

Every possible facility and control

PANAFLEX cameras have in-shot adjustable shutters, behind the lens filters, full-fitting register pins for "plate steady" image steadiness, pitch control to set the pull-down stroke for maximum quietness, crystal controlled speeds, variable speeds from 4-36 f.p.s., the possibility of single shot and slow running.

On the PLATINUM PANAFLEX it is possible to use the computer-measured camera shutter opening and speed control systems to shoot with HMI lights at any camera speed, enabling speeded up and slowed down action to be shot without fear of light flicker.

The adjustable shutter system of all PANAFLEX cameras, combined with the wide range of electronic synchronizing units, makes it possible to shoot with HMI, and other discharge lamps, in absolute safety; to film TV screens at 24 or 25 f.p.s. with 144, 172.8 or 180° shutter opening; and to synchronize with process projectors and every possible TV scanning speed.

When you "go PANAVISION" a full palette of additional possibilities becomes available.

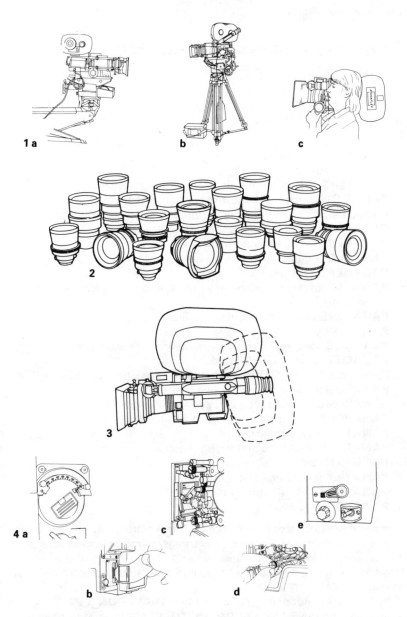

1.a., b. & c. PANAFLEX cameras may be dolly or tripod mounted, or hand-held,
2. PANAVISION PRIMO lenses, the latest additions to the wide range of spherical
and anamorphic lenses, 3. PANAFLEX film magazines may be mounted on top
or at the rear of the camera, 4. PANAFLEX cameras have: a. an "in-shot"
adjustable shutter, b. behind-the-lens filtering, c. full-fitting register pins,
d. adjustable pitch control and e. integral variable speed control

45

Panaccessories to Help the Cinematographer

Every accessory for PANAFLEX cameras, many of them unique and exclusive to PANAVISION, are thought through as an integrated part of the system:

PANABALL leveller/hi-hat for any Mitchell-type tripod head
PANACLEAR heated mist-free eyepiece
PANAFLASHER in-camera negative flashing device
PANAFADE electronic exposure control system
PANAFINDER hand-holdable viewfinder which uses the camera lens
PANAFRAME video-assist freeze-frame device
PANAGLIDE floating camera systems
PANAGLOW illuminated ground-glass reticle
PANAHEAD pan and tilt head
PANALAB advanced technology optical design and lens testing department
PANALENS LIGHT lens calibration illuminating light
PANALEVELLER eyepiece-levelling device
PANALITE constant color temperature/variable intensity Obie lamp
PANALUX nightvision viewfinder
PANANODE ADAPTOR for nodal panning and tilting on a PANAHEAD
PANAPOD lightweight yet rugged tripod
PANAREMOTE remote control system for any PANAHEAD
PANAROCK near ground level pan and tilt device
PANATAR anamorphic lenses
PANATAPE electronic measuring devices
PANATATE 360° nodal turnover mount
PANATILT tilt/balance plate
PANAVID TV viewfinder systems
PANAZOOM zoom lenses

Additional optional accessories include frame cutters, car rigs for rugged tracking, rain deflectors, underwater housings, splash boxes, weather proof covers, remote focus, zoom, aperture and shutter controls, camera and lens heaters, multi-stage matte boxes and companion cameras which accept all PANAVISION lenses.

As if all that is not enough there is the **PANAVISION 3-PERF** camera movement and the **PANAVISION AATON CODE** time code system.

PANABALL

PANACLEAR

PANAFADE

PANAFINDER

PANAFLASHER

PANAFRAME

PANAGLIDE

PANAGLOW

PANAHEAD

PANALAB

PANALENS LIGHT

PANALEVELLER

PANALITE

PANALUX

PANANODE

PANAPOD

PANAREMOTE

PANAROCK

PANATAPE

PANATAR LENSES

PANATATE

PANATILT

PANAVID

PANAZOOM

47

Light Efficiency

With a maximum shutter opening of 200° and a wide range of ultra-wide aperture lenses ranging from T1.0 and T1.3 (Spherical) and T1.1 and T1.4 (Anamorphic) PANAFLEX cameras are the most "light-efficient" cameras available. At T1.0 and rating the film at EI 500 less than 1 foot candle of light is necessary to give full exposure.

The 200° shutter passes ⅙ stop more light than a 175-180° shutter and only requires to be closed down to 100° to reduce the exposure by one full stop and to 50° to reduce one stop again. The minimum shutter opening on the PANAFLEX is 50°.

In addition to transmitting more light, the PANAFLEX 200° shutter opening reduces the possibility of image strobe during fast pan movements.

Sharper and steadier images

PANAFLEX cameras have a focal plane shutter which moves across the shortest side of the frame and in the opposite direction to the travel of the film. This adds very considerably to the camera's light capping efficiency. Very precise shutter timing ensures that all light is completely cut off before the film begins to move and remains so until it has come to a stop and has been held securely in position by the registration pins. Unlike many other cameras there is absolutely no movement of the film while even the smallest part of the frame is uncovered, even at high speed.

Another important point is that PANAFLEX registration pins are full fitting and no compromise is made to quiet the camera. Furthermore, the pins are positioned on both sides of the film and engage in the perforations immediately below the frame line, the same as on all optical printers.

You can safely use any PANAFLEX camera to shoot plates for multi-exposure SFX scenes.

The PANALUX viewfinder

To make it easier to see picture detail through the viewfinder when setting up and operating in extremely low light levels the regular viewfinder extension tube may be exchanged for a PANALUX nightsight type which intensifies the image more than 1000 times.

You can see more with PANAVISION.

EXPOSURE CONTROL BY SHUTTER ADJUSTMENT

EXPOSURE	SHUTTER OPENING (°)	
Full open	200	180
− ⅓ stop	159	143
− ⅔ stop	126	114
− 1 stop	100	90
− 1⅓ stop	80	72
− 1⅔ stop	63	57
− 2 stop	50	45

180°-200° EXPOSURE ADJUSTMENT

200° EXPOSURE SETTING FOR 180° EXPOSURE READING									
180°	1	1.4	2	2.8	4	5.6	8	11	16	22
200°	1.05	1.45	2.1	2.9	4.2	5.9	8.4	11.5	16.8	23.1

CAMERA SPEED/SHUTTER ANGLE EXPOSURE TIMES

Shutter angle (°)	CAMERA SPEED (fps)								
	6	8	10	12	16	20	24	25	30
	EXPOSURE TIME (secs.)								
40	0.0185	0.0139	0.0111	0.0093	0.0069	0.0056	0.0046	0.0044	0.0037
60	0.0278	0.0208	0.0167	0.0139	0.0104	0.0083	0.0069	0.0067	0.0056
80	0.037	0.0278	0.0222	0.0185	0.0139	0.0111	0.0093	0.0089	0.0074
100	0.0463	0.0347	0.0278	0.0231	0.0174	0.0139	0.0116	0.0111	0.0093
120	0.0556	0.0417	0.0333	0.0278	0.0208	0.0167	0.0139	0.0133	0.0111
140	0.0648	0.0486	0.0389	0.0324	0.0243	0.0194	0.0162	0.0156	0.013
144	0.0667	0.05	0.04	0.0333	0.025	0.02	0.0167	0.016	0.0133
160	0.0741	0.0556	0.0444	0.037	0.0278	0.0222	0.0185	0.0178	0.0148
172.8	0.08	0.06	0.048	0.04	0.03	0.024	0.02	0.0192	0.16
180	0.0833	0.0625	0.05	0.0417	0.0313	0.025	0.0208	0.02	0.0167
200	0.0926	0.0694	0.0556	0.0463	0.0347	0.0278	0.0231	0.0222	0.0185

The art of cinematography is in image control.

Optical Accessories

There is more to creative cinematography than setting the lights and getting the exposure correct. PANAVISION has developed a whole series of accessories to put the cinematographer's imagination on the screen.

Filters
PANAVISION can supply the widest possible range of image control filters and "clever" matte boxes to put them in. To keep the number of filters required to cover their lenses to a minimum most PANAVISION lenses are covered by the standard size, 4 × 5.650", filters. However, there are occasions when a cameraman wishes to slide and rotate filters, especially graduated filters, and to accommodate this PANAVISION also supplies multi-stage matte boxes which take 6.6" sq. filters.

PANAVISION tests its filters on its optical test bench to ensure that their optical quality is every bit as good as its lenses.

Even the 10mm wide-angle PRIMO lens has a specially designed clip-on filter holder to utilize standard 6.6" sq. filters, thus eliminating the need to handle exceptionally large, awkward and additional filters.

Diopters
A very useful accessory, especially with zoom lenses, are full-cover "diopter" lenses, which are placed in front of the normal taking lens for closer focusing. PANAVISION is able to supply diopter lenses to cover all their zoom lenses.

Split diopters are another very useful accessory, especially for Anamorphic cinematography, as these make it possible to have two planes of focus in a single shot. Split diopters are full diopters cut in half and are placed in front of the normal lens so that the close-up part of the scene is photographed through the diopter.

When using split diopters it is usual to lose the dividing line between the two planes of focus by placing the edge of the diopter coincident with a vertical object such as a pillar or the corner of a room.

Diffusion
PANAVISION has a wide range of standard and customized sliding diffusers available to enable Directors of Photography to control the amount of diffusion as a scene progresses, usually as they pan to, from and between close-ups where one artist requires more diffusion than another.

Special matte boxes
PANAVISION can supply matte boxes with filter trays which slide, rotate and tilt to make the most use of a particular type of filter.

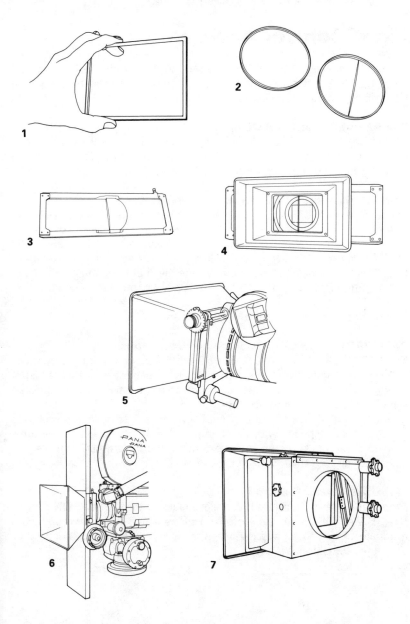

1. Standard 4 × 5.650″ filter which covers most PANAVISION lenses, 2. Full cover and split diopters for zoom lenses, 3. Split diopter mounted on a slide, 4. Sliding split diopter mounted in front of an anamorphic lens, 5. Standard sliding diffuser fitted to the rear of a standard matte box, 6. Custom-made sliding diffuser, 7. Matte box fitted with a tiltable filter stage.

Lighting is the very bloodstream of cinematography.

Light Control Accessories

The ways to make PANAFLEX cameras interface with the lighting add much not only to the way the image photographs, but also to the way in which the camera can be operated.

The PANALITE onboard Obie light
The PANALITE is a shadowless fill-light which fits onto the camera just above the matte-box and which can be dimmed to almost nothing without affecting the color temperature of the light.

PANAFLEX micro precision adjustable shutter
The in-shot ADJUSTABLE FOCAL PLANE SHUTTER has always been one of the most important of all PANAFLEX features. The advent of the on-board micro computer has enabled the ADJUSTABLE SHUTTER of the PLATINUM PANAFLEX to be controlled to greater accuracy than ever before. This has made it possible to fine tune the shutter opening to suit any HMI Lighting Power Supply Hz/Camera Speed combination without fear of HMI flicker.

The effective exposure time = (shutter angle/360) $\times$ (1/fps).

PANAFADE electronic exposure control system
The PANAFADE electronic EXPOSURE CONTROL unit system comprises two counter-rotating Polaroid filters which reduce (or increase) the light transmission by two stops as they are rotated relative to one another. (The initial light loss is three stops.)

Originally developed for the PANASTAR camera, to make it possible to change camera speeds without changing the characteristics of the lens, as would happen if a stop pull were done, the PANAFADE can also be used with all the latest PANAFLEX cameras.

Motorized graduated filters
Among the "clever" facilities that are available for the PANAVISION MULTI-STAGE MATTEBOX are the motorized slides which enable two opposing graduated filters to be slid towards (or away from) one another to give an overall transitional effect.

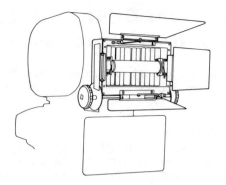

1. PANALITE on-board Obie light

2. HMI FRAME PER SECOND/SHUTTER ANGLE TABLES

OPTIMUM SHUTTER ANGLE (°) FOR HMI LIGHTING					
6-20 fps					
fps	6	8	10	12	16	20
60Hz	198*	192*	180*	180*	192	180*
50Hz	194.4	172.8	180*	172.8	172.8	144
22-36 fps					
fps	22	**24**	25	26	30	36
60Hz	198*	**144***	150	156	108*	108
50Hz	158.4	**172.8**	180*	187.2	108	129.6

*This is an optimum shutter opening but any shutter opening is possible if
the f.p.s. and the Hz are precise.

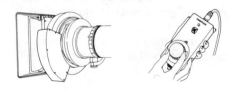

3. PANAFADE unit

53

Light can be filtered or applied on either side of the lens.

In-Camera Light Control and Enhancement Accessories

In addition to offering the means to control the light entering the camera, PANAFLEX cameras also provide facilities to manipulate the light inside the camera.

Behind the lens filters

There are many advantages to placing at least one of the light control filters (usually the 85) behind the lens; for one, it makes room for one more fog, diffuser, low-contrast or other filter in the matte box. Behind the lens filtering has always been a feature of all PANAFLEX and PANASTAR cameras.

PANAFLASHER light overlay accessory

Since the very beginnings of photography film has been "flashed" with a very low brightness light either before or after exposure, to reduce contrast or increase exposure in the shadow areas, or to put a color bias into the darker tones without greatly affecting the highlights.

In the past it was a process normally carried out by a film laboratory at a price per foot of film and with no scene-by-scene control by the cinematographer.

The effect is to give the film a basic exposure below the toe of the sensitometric curve. This overcomes the exposure inertia so that all light falling on the film during a take appears on the screen.

The PANAVISION PANAFLASHER fits onto whichever PANAFLEX magazine port is not in use. (If the magazine is on the top of the camera the PANAFLASHER fits on the rear and vice versa.)

The PANAFLASHER unit is small enough to be used while the camera is hand-held and does not require a main power source.

The PANAFLASHER incorporates an extremely sensitive exposure meter so that the effect can be very closely monitored and controlled.

White light will control image contrast and bring out details in shadows that would otherwise be plain black. Colored light will tint the darker parts of the image leaving the highlights untouched and having very little effect on light skin tones.

PANAVISION can also supply the necessary interface units to fit a Light-flex light overlay system to any PANAFLEX camera.

54

1 a

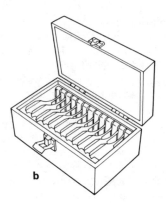

b

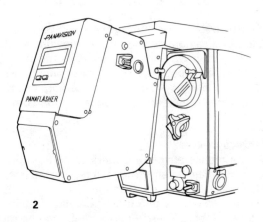

2

1.a. Behind the lens gelatine filter holder and b. box of filters, 2. PANAFLASHER unit fitted to the rear of a PANAFLEX camera.

Camera Angle Control Accessories

Not every shot has to be taken with the camera at eye level and perfectly horizontal. With PANAVISION it is possible to have fun with camera angles.

Inclining prisms
An INCLINING PRISM placed in front of the taking lens enables either very low or very high angle shots which would otherwise be impossible.

An INCLINING PRISM works by reflecting and refracting the light and because the light is bent twice the image is the correct way round. Other optical advantages are that there are only two air-to-glass surfaces so there is virtually no loss of transmission. The planes of the front and rear of the prism are perfectly flat so there is no chance of optical distortion, and by using a glass of high refractive index it is possible to cover a much wider angle lens than would otherwise be possible.

PANAGLIDE floating camera systems
The possibility of taking the camera off the tripod, or other rigid support system, and moving it about with all the fluidity of human movement is a feature of modern cinematography brought about by the floating camera systems. The advantage of these systems is that they make unobtrusive gliding camera movements possible without the audience being aware.

PANAVISION can supply PANAGLIDE floating camera systems complete or can supply a lightweight PANAFLEX to use with Steadicam rigs.

PANATATE turnover camera mounts
PANATATE turnover camera mounts are just about the most refined mounts of that sort ever designed. They not only have the capability to rotate a camera longitudinally (about the lens axis) during a take but also incorporate a nodal mounting system so that a camera can be panned, tilted and rotated simultaneously without displacing the entrance pupil of the lens.

LOUMA camera cranes
LOUMA camera cranes enable a camera to be positioned in otherwise inaccessible places while maintaining full facilities to pan, tilt, dolly, raise it up and down, and control the lens focus, zoom and aperture settings as though it were on a dolly or a truck-mounted crane with all the crew.

LOUMA camera cranes liberate the camera from the restrictions of human accessibility and have advanced the art of creative film making in a most positive manner.

Not least of the possibilities that LOUMA camera cranes open up is the ability to place a camera in dangerous positions, such as over the edge of mountains or tall buildings, or to dip a camera into a set, do a 360° pan and then elevate the camera from eye level to 40' high or more.

56

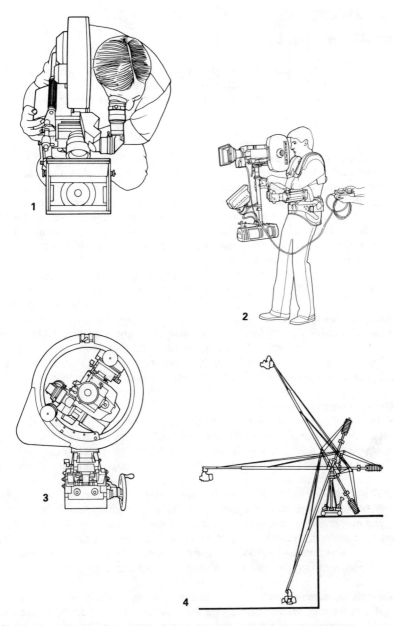

1. Inclining prism low-angle attachment fitted to a hand-held PANAFLEX camera for a ground-level shot, 2. PANAGLIDE floating camera system, 3. PANATATE turnover camera mount, 4. LOUMA camera crane.

Environmental Protective Equipment

Creative cinematography often demands that a camera must work perfectly under conditions of intense cold, desert heat and dust, driving rain, sea and storm, severe vibration, stress and "g" forces and in dangerous places.

Intense cold
PANAFLEX cameras and magazines are fitted with internal heaters which ensure that they run freely and quietly under even the coldest conditions.

Additional heater barneys are available to cover and give additional warmth to the camera, the magazines and the lenses (especially zoom lenses) and are advisable for use when the camera is set up in an exposed position where the wind chill factor is likely to be significant. Where camera heaters and heater barneys are likely to be used intensively, additional camera batteries should be ordered.

Lenses kept in a cold truck and taken into a warm building for use should be stored in polythene bags to prevent internal misting up.

Heat and dust
Thanks to their light color, smooth finish and good sealing PANAFLEX cameras operate exceptionally well in conditions of extreme heat and dust.

Dust should be brushed away from the camera, not cleared with an aerosol air spray which only blows it into inaccessible places.

Rain, storm and water
Waterproof covers are available to protect the camera from rain, etc.

Spinning disk spray deflectors/waterproof covers are available for use in storm conditions. These devices incorporate high speed rotating glass disk fronts which throw off water as quickly as it falls on them, thus keeping lenses clear of water.

Water boxes are available for shallow water filming (up to 1' deep) and are useful for surface filming in water tanks, swimming pools, etc.

If a camera is contaminated with salt water all traces of salt should be washed away with fresh water at the earliest possible opportunity and the camera returned to PANAVISION, or its representative, for emergency servicing. Nothing can damage a camera more than salt water.

Vibration, stress, high "g" forces and dangerous places
Automobile mounts and clamp rigs are available to give additional camera and magazine security in rugged conditions, as when filming car chases.

Cameras in protective boxes and "disposable" cameras are available for use in highly hazardous positions.

58

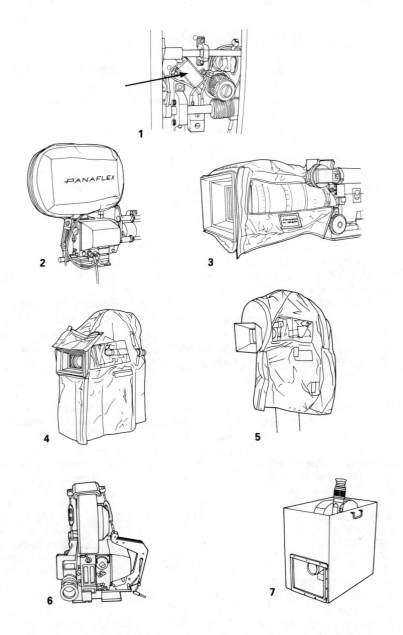

1. Internal Heater elements arranged around a camera bearing, 2. Magazine heater/cover showing power supply arrangements, 3. Zoom lens heater/cover, 4. Normal waterproof camera cover, 5. Rain deflector waterproof cover, 6. Camera clamp rig, 7. Water-box shallow water housing.

59

PANAVISION People

Of special note are the dedicated PANAVISION PEOPLE who use all their experience to prepare the equipment for your shoot with all the efficiency and enthusiasm it is possible to muster. They like to be considered to be a part of your extended crew.

They will gather all the equipment together for you and your crew to make the most exacting tests you can devise. They will listen to any comments you have to make and change or modify anything with which you are not entirely happy. They will want to hear from you while you are away and when you return. In particular they will want to know if anything at all went wrong.

They will welcome suggestions as to what can be done to improve and extend the PANAVISION product range even further. PANAVISION's policy of constant updating ensures that today's new idea will be incorporated into every existing camera.

If the PANAFLEX range is not already your ideal camera system they would even like to know what else needs to be done to make it so.

The PANAVISION PEOPLE are your constant partners in progressive cinematography and every time a Director of Photography wins or is nominated for an Oscar using PANAFLEX cameras and lenses which they have conceived, designed, manufactured and prepared, they quietly share in your personal pleasure and satisfaction.

Cinematographers who have won Oscars and Oscar Nominations for Best Cinematography using PANAVISION cameras and/or lenses (1958-88):
(Winners in bold)

NESTOR ALMENDROS	**Days of Heaven**
NESTOR ALMENDROS	Kramer vs. Kramer
NESTOR ALMENDROS	The Blue Lagoon
NESTOR ALMENDROS	Sophie's Choice
JOHN A. ALONZO	Chinatown
JOSEPH BIROC	**The Towering Inferno**
PETER BIZIOU	**Mississippi Burning**
RALPH D. BODE	Coal Miner's Daughter
BILL BUTLER	One Flew Over the Cuckoo's Nest
GHISLAIN CLOQUET	**Tess**
JACK COUFFER	Jonathan Livingstone Seagull
JAMES CRABE	The Formula
JORDAN CRONENWETH	Peggy Sue Got Married
DEAN CUNDEY	Who Framed Roger Rabbit

60

ALLEN DAVIAU	E.T. the Extra-Terrestrial
ALLEN DAVIAU	The Color Purple
ALLEN DAVIAU	Empire of the Sun
ERNEST DAY	A Passage to India
CALEB DESCHANEL	Wargames
CALEB DESCHANEL	The Natural
DANIEL L. FAPP	Ice Station Zebra
DANIEL L. FAPP	**West Side Story**
WILLIAM A. FRAKER	Looking for Mr. Goodbar
WILLIAM A. FRAKER	Heaven Can Wait
WILLIAM A. FRAKER	1941
WILLIAM A. FRAKER	Murphy's Romance
OSAMI FURUYA	Tora! Tora! Tora!
LEE GARMES	The Big Fisherman
CONRAD HALL	**Butch Cassidy and the Sundance Kid**
CONRAD HALL	The Day of the Locust
CONRAD HALL	Tequila Sunrise
SINSAKU HIMEDA	Tora! Tora! Tora!
RICHARD H. KLINE	King Kong
FRED KOENEKAMP	Patton
FRED KOENEKAMP	**The Towering Inferno**
JOSEPH LaSHELLE	The Apartment
ERNEST LASZLO	It's a Mad, Mad, Mad, Mad World
ERNEST LASZLO	Logan's Run
PHILIP LATHROP	Earthquake
SAM LEAVITT	Exodus
JOSEPH MacDONALD	The Sand Pebbles
OSWALD MORRIS	Oliver!
OSWALD MORRIS	**Fiddler on the Roof**
ASAKAZU NAKAI	Ran
SVEN NYKVIST	The Unbearable Lightness of Being
MIROSLAV ONDRICEK	Amadeus
DON PETERMAN	Flashdance
DON PETERMAN	Star Trek IV: The Voyage Home
OWEN ROIZMAN	The Exorcist
OWEN ROIZMAN	Network
OWEN ROIZMAN	Tootsie
TAKAO SAITO	Ran
MESAMICHI SATOH	Tora! Tora! Tora!
JOHN SEALE	Witness
JOHN SEALE	Rain Man
DOUGLAS SLOCOMBE	Travels with my Aunt
DOUGLAS SLOCOMBE	Julia
DOUGLAS SLOCOMBE	Raiders of the Lost Ark
HAROLD E. STINE	The Poseidon Adventure
HARRY STRADLING	**My Fair Lady**
HARRY STRADLING Jr.	1776
HARRY STRADLING Jr.	The Way We Were

Cinematographers who have won Oscars and Oscar Nominations for Best Cinematography using PANAVISION cameras and/or lenses, (continued):

ROBERT SURTEES	**Ben Hur**
ROBERT SURTEES	Mutiny on the Bounty
ROBERT SURTEES	The Last Picture Show
ROBERT SURTEES	Summer of '42
ROBERT SURTEES	The Sting
ROBERT SURTEES	The Hindenburg
ROBERT SURTEES	A Star is Born
ROBERT SURTEES	The Turning Point
RONNIE TAYLOR	**Gandhi**
MASAHARU UEDA	Ran
GEOFFREY UNSWORTH	Murder on the Orient Express
GEOFFREY UNSWORTH	**Tess**
HASKELL WEXLER	One Flew Over the Cuckoo's Nest
HASKELL WEXLER	**Bound for Glory**
CHARLES F. WHEELER	Tora! Tora! Tora!
BILLY WILLIAMS	On Golden Pond
BILLY WILLIAMS	**Gandhi**
GORDON WILLIS	Zelig
JAMES WONG HOWE	**Hud**
JAMES WONG HOWE	Funny Lady
FREDDIE YOUNG	**Lawrence of Arabia**
FREDDIE YOUNG	**Doctor Zhivago**
FREDDIE YOUNG	**Ryan's Daughter**
FREDDIE YOUNG	Nicholas and Alexandra
VILMOS ZSIGMOND	**Close Encounters of the Third Kind**
VILMOS ZSIGMOND	The Deer Hunter
VILMOS ZSIGMOND	The River

. . . and more to come

The
Camera Operators'
PANAFLEX

The Operator has the closest of all relationships with the camera.

The Camera Operators' PANAFLEX

The aspects of camera design that affect the Camera Operator most—the control of pan and tilt movements, the large, comfortable, eyepiece and the quality of the viewfinder system, and the fact that the film magazines may be fitted on top or at the rear, which together with the ergonomic shape, the good balance and the comfort pads makes hand-holding almost effortless—are all areas where the PANAFLEX camera system is particularly superior.

PANAHEAD facilities

With PANAVISION Operators have the choice of two PANAHEADS, the "Regular" model, which is suitable for most PANAFLEX camera and lens combinations, and the "Super" version, which is preferable for PANAFLEX and PSR cameras in combination with particularly heavy lenses.

The PANAHEAD pan and tilt camera head derives its unique smoothness of movement from the patented drive system which translates the smallest hand movement to a camera movement without the problems of gear cogging and wear in one section of the tilt quadrant. These are unavoidable problems inherent in all traditional geared heads with metal quadrants.

The PANAHEAD offers the Camera Operator a choice of three pan and tilt speeds in the "geared head" mode. It may also be used as a free head, as a gyro head and as a remote head.

With a PANAHEAD it is possible to film at any angle, from 90° directly up to 90° directly down, encompassing the widest possible range of tilt.

PANAFLEX viewfinder facilities

The "Brighter than Life" viewfinders of PANAFLEX cameras give the brightest ground-glass images of any film camera. For most models there is a choice of three viewfinder lengths: short for hand-holding, intermediate for getting close to the camera when it is on a head, and long for standing back from the camera. On the PANAFLEX-16 the viewfinder can be swung out for left eye viewing.

Other much appreciated viewfinding features are the patented PAN-AGLOW system, which lights up the viewfinder markings when shooting against a dark background, the patented PANALEVELLER, which keeps the eyepiece at eye level irrespective of how much the camera is tilted up or down, the PANACLEAR heated eyepiece, which positively prevents the viewfinder optics from misting up in cold conditions, the ocular marker ring, image magnification, a choice of two contrast filters and a de-anamorphoser, which makes the image larger, not smaller.

All PANAFLEX cameras have facilities for fitting video-assist systems of exceptional sensitivity, which may be flicker-free if required.

On the PLATINUM PANAFLEX there is even a place for the Operator, or anyone else, to park his or her spectacles while looking through the viewfinder!

64

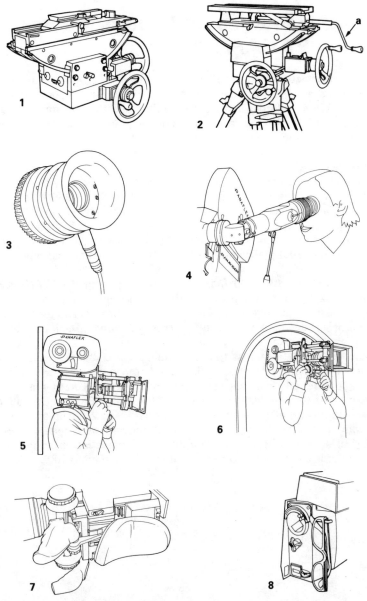

1. Regular PANAHEAD, 2. Super PANAHEAD with adjustable tilt (a.), 3. The large, comfortable, PANAFLEX eyepiece, 4. The swing-out viewfinder facility of the PANAFLEX-16, 5. Hand-holding a PANAFLEX in a confined space, 6. Hand-holding a PANAFLEX with little headroom, 7. The PANAFLEX shoulder pad and ergonomic handgrips for comfortable handholdability, 8. The PLATINUM PANAFLEX spectacle park.

65

Sharper Images and Faster Pans with a PANAFLEX

Of the many features which differentiate the PANAFLEX camera system from others are the light capping efficiency of the local plane shutter, which makes for sharper images on the film, and the maximum shutter opening of 200°, which maximizes the exposure time and makes it possible to pan faster without fear of strobing.

Panning speeds

Camera Operators should be aware that due to the efficiency of the focal plane shutter of the PANAFLEX camera and the fact that the registration pins do not disengage until the film is completely capped, nor does the shutter begin to open until the film is firmly held in position once more, the images on the film tend to be crisper and sharper than on cameras which are not so light-efficient. This may affect the maximum safe panning speed.

On the other hand, the 200° maximum shutter opening of the PANAFLEX camera maximizes the image blur on the leading and trailing edges of the image, making it possible to pan faster than with a smaller shutter opening. The vertical movement of the shutter across the film makes it possible to pan in either direction equally fast.

How fast is a safe "across frame" panning speed (the time it takes for a fixed object to travel from one side of the frame to the other, irrespective of lens focal length) is a subjective judgement which Operators take note of every time they see dailies. It is part of the skill of Camera Operating.

As a starting point, most Operators consider that a 24/25 f.p.s. and with the shutter at 180° a 5 sec. "across frame" movement is quite safe with spherical lenses, 7 secs. with anamorphic. These times will be longer when the shutter is closed down, less when it is fully opened to 200°.

CAMERA SPEED/MINIMUM 'ACROSS FRAME' PAN TIME TABLE
(Minimum safe 'across frame' pan time with a normal shutter angle but with a modified camera speed)

Modified camera speed (fps)	Normal 24fps/180° safe 'across frame' pan time (sec)					
	4	5	6	7	8	10
	New minimum safe 'across frame' pan time (sec)					
6	16	20	24	28	32	40
8	12	15	18	21	24	30
10	10	12	14	17	19	24
12	8	10	12	14	16	20
16	6	8	9	11	12	15
20	5	6	7	8	10	12
24	4	5	6	7	8	10
30	3	4	5	6	6	8
40	2	3	4	4	5	6
60	2	2	2	3	3	4
80	1	2	2	2	2	3
100	1	1	1	2	2	2
120	1	1	1	1	2	2

CAMERA SPEED/MINIMUM SHUTTER ANGLE TABLE
(To pan at the normal 'across frame' pan time but with a modified camera speed, using the adjustable shutter to compensate)

Modified camera speed (fps)	6	8	10	12	16	20	24	25	26.6
New minimum shutter angle (°)	45	60	75	90	120	150	180	188	200

SHUTTER ANGLE/MINIMUM SAFE 'ACROSS FRAME' PAN TIME TABLE
(Minimum safe 'across frame' pan time at normal camera speed but with a modified shutter angle)

Modified shutter angle (°)	Normal 24fps/180° safe 'across frame' pan time (secs)					
	4	5	6	7	8	10
	New safe 'across frame' pan time (sec)					
40	18	23	27	32	36	45
60	12	15	18	21	24	30
80	9	11	14	16	18	23
100	7	9	12	13	14	18
120	6	8	9	11	12	15
140	5	6	8	9	10	13
160	5	6	7	8	9	11
180	4	5	6	7	8	10
200	4	5	5	6	7	9

Note: All the above figures are rounded off to the nearest whole number for the sake of simplicity.

PANAVID Viewfinder Systems

PANAVISION has always been at the forefront in developing video-assist systems and embracing the most advanced video technologies as they emerge. The various systems they offer, the PANAVISION CCD FLICKER-FREE PANAVID, the SUPER PANAVID and the regular PANAVID each represent the "state of the art" at the time they were developed.

PANAVISION has concentrated on developing video assist systems with the best possible resolution, tonal gradation and sensitivity, avoiding color until it is possible to provide a system that is not only as good as the current B & W systems in these respects but also renders color in a way that is not widely out of balance with what is to be expected on film. Be sure that when that day comes PANAVISION will have a color PANAVID.

The CCD flicker-free PANAVID

The PANAVISION FLICKER-FREE PANAVID video assist system uses a 740 × 488 pixel high-resolution CCD camera which gives a completely flicker-free video image at any camera speed from 4 to 120 f.p.s. This makes it much more pleasant for the Operators to view and gives a truer rendition of light and shade. The extreme sensitivity of the system and the automatic gain control make it possible for the Operator to see action in dark sections of a scene, yet the image does not burn out when confronted with a bright light source.

For normal lighting conditions the PANAVID CCD incorporates an infrared filter to ensure best balance color rendition across the full grey scale but for low-light conditions this filter may be removed. Conversely, a neutral density filter may be interposed in the CCD light path to reduce the amount of light.

Image grabbing

The FREEZE-FRAME facility makes it possible to grab a single frame from a scene, just like taking a Polaroid still. It can also be held in store and used as an electronic matte or cut frame to be aligned with another setup in the future.

Equally, a frame can be imported from an outside source so that a current setup can be accurately aligned with a previous scene.

When shooting Traveling Matte scenes it is possible to overlay the current image onto a previously shot background plate in exactly the same manner as it will appear in composite scene.

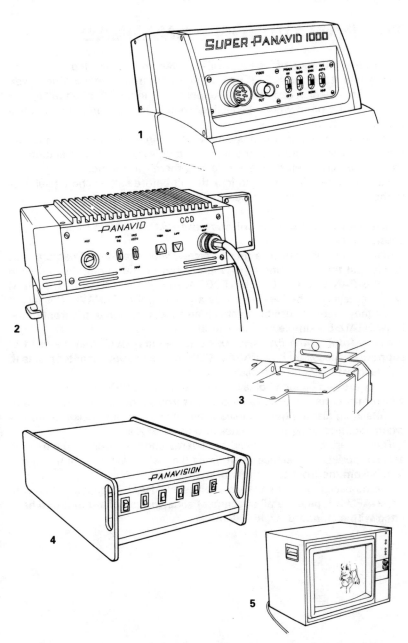

1. SUPER PANAVID 1000 video-assist camera, 2. PANAVID flicker-free CCD video-assist camera, 3. PANAVID CCD filter wheel, 4. PANAVID PANAFRAME control box, 5. Illustration of aligned images

69

What tilts up must tilt down. . . . just as smoothly.

The PANAVISION PANAHEAD

The normal tilt range of the PANAHEAD is 30° up or down. It goes without saying that it can do a 360° pan around and around many times over. Three handwheel speed ratios make possible the full 30° of tilt in 15, 8 or 4 turns of the hand wheel respectively and a 360° pan in approximately 75, 41 or 21 turns.

The handwheel finger knobs can be preset in a precise position to suit a specific shot by returning the ratio selector lever to "N" and setting the knob as required before finally selecting the desired ratio.

Note: It is very important to lock the tilt before setting the tilt selector to the "N" position, or even passing through it.

Additional features that enhance the PANAHEAD's user friendliness

Where to put the exposure meters safely, keep the rolls of camera tape handy, put the measuring tape where it can be found quickly, and even put the PANAVISION CINEMATOGRAPHERS' COMPUTER PROGRAM where an eye can be kept on it was a problem that PANAVISION solved when they made it possible to attach an accessory box to the front of the PANAHEAD. So simple and so useful!

At the front of the PANAHEAD are two studs which may be used to support an accessory box. PANAVISION Inc. can supply adaptor plates to interface with clients' own accessory boxes.

There are two bushed holes on either side of the PANAHEAD tilt section which may either be used to pass a bar through for carrying purposes or for attaching lashing down cables when maximum rigidity is required as when shooting background plates and other SFX shots.

The camera body may be slid backwards and forwards on the head to achieve perfect tilt balance at all times (i.e., when exposed film is transferred from the front to the rear of the magazine).

Sliding base plates are available to use a PANAFLEX camera fitted with a PANAFLEX type dovetail attachment slide with any flat-top tripod head fitted with a standard ⅜" 16 TPI. thread.

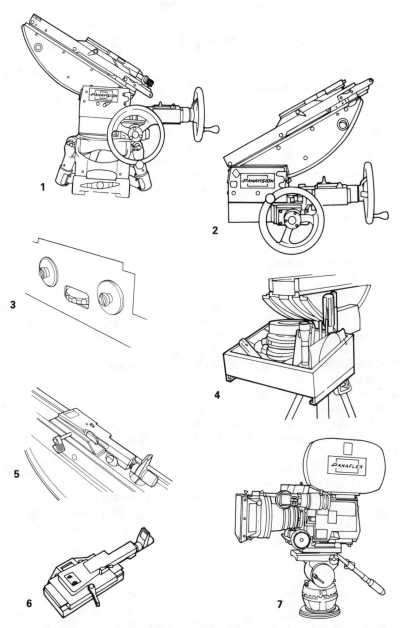

1. PANAHEAD tilted 30° up, 2. PANAHEAD tilted 30° down, 3. Accessory box studs on the front of a PANAHEAD, 4. Assistant's accessory box fitted to the front of a PANAHEAD, 5. Sliding base unit on top of a PANAHEAD, 6. Separate sliding base unit for use with fluid heads, etc., 7. PANAFLEX camera mounted on a flat-top type tripod head.

71

It would be impossible to have a greater tilt range.

Additional Camera Tilt, Up or Down

An adjustable tilt plate for exaggerated tilt up or down makes possible shots with the camera pointing 90° up or 90° down.

To adjust the tilt plate for additional tilt down
For additional tilt DOWN release the rear tilt plate locking lever that is just in front of the rear pan handle socket and press the safety catch at the rear of the tilt plate on the left. For small amounts of additional tilt, press in whichever of the three safety stops at the rear right hand side of the dovetail slideway is appropriate and lock the sliding block in such a manner that the safety stop remains in position. Greater amounts of tilt may be achieved by additionally pressing one of the three safety stops at the front and adjusting the forward block accordingly.

To adjust the tilt plate for additional tilt up
For additional tilt up loosen both the front and the rear tilt plate locking levers, depress either the front or back lower dovetail slide safety stops to release, slide out and reverse the entire tilt plate subassembly, reverse the camera attachment plate so that the entire subassembly operates in the opposite direction and proceed as for additional tilt DOWN.

The PANAROCK and the PANATILT accessories
For additional tilt at ground level or close to any flat surface the PANAROCK and the PANATILT are two very useful accessories.

The PANAROCK is a semicircular device which fits on the underside of a PANAFLEX camera and enables the camera to be positioned very close to the floor or other flat surface and tilted up and down in shot by a rocking motion.

The PANATILT is an adjustable wedge unit which may also be fitted to the underside of a PANAFLEX camera to give additional tilt movement in close proximity to a flat surface. The PANATILT unit provides a more rigid support than the PANAROCK and is not suitable for in-shot camera movements.

72

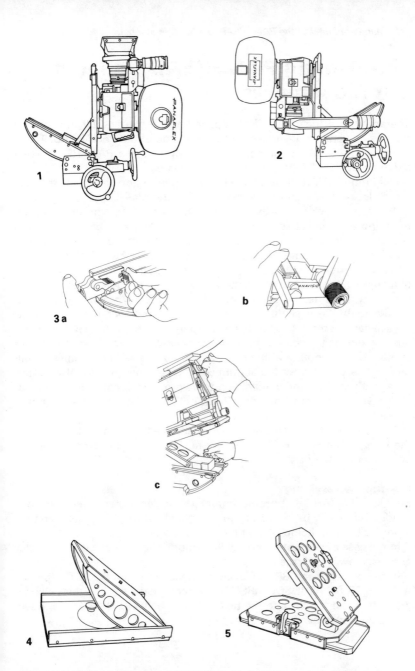

1. PANAHEAD tilted 90° up, 2. PANAHEAD tilted 90° down, 3. Adjusting a PANAHEAD for additional tilt up or down: a. release the side locking lever, b. press in the safety catch, c. raise the dovetail slide and securely lock off, 4. PANAROCK, 5. PANATILT.

Fine Tuning PANAHEAD Pan and Tilt Movements

The amount of friction or "feel" applied to the pan and tilt movements may be adjusted by levers at the rear and side of the PANAHEAD. These levers may be fully tightened to positively lock the PANAHEAD when shooting Special Effects and background plates.

Later model PANAHEADS have tilt locking levers on either side to make the tilt lock-off even more positive for shooting plates.

Any gross backlash in the pan movement may be smoothed out by releasing a small set screw at the rear right hand side of the head and adjusting a knob at the rear. Take care not to set the gears so tight that they bind. This adjustment is correctly set before a PANAHEAD leaves the PANAVISION plant or that of any of its representatives, but a small amount of adjustment may be required during the course of a long location shoot, especially if the head is used constantly with the keyway in one position relative to the scene so that gear wear is maximized over a short segment.

Take care also not to completely undo the pan adjusting knob and disengage the pan gears as these are very meticulously lapped together relative to one another at the factory. Should this setting be upset it will be necessary to remove the bottom cover plate from the PANAHEAD and note that the small index marks on the worm wheel and drive gears coincide when they come together. This is not a job that should be attempted in the field.

The tension of the belt drive may be adjusted by turning the wheel at the front of the PANAHEAD to the right. This also should rarely require adjustment in the field.

Emergency servicing

Should a PANAHEAD become immersed in sea water or subject to a sandstorm, it will be necessary to completely strip down, clean and relubricate the unit before it is fit for further use.

In the case of sea water immediately submerge the head in fresh water and wash away all traces of salt.

In the case of sand BRUSH away all traces of sand. DO NOT USE AN AIRLINE OR AN AEROSOL AIRSPRAY to remove the sand as this will only drive the sand further into the bearings and between the gears and will inevitably cause even more damage.

74

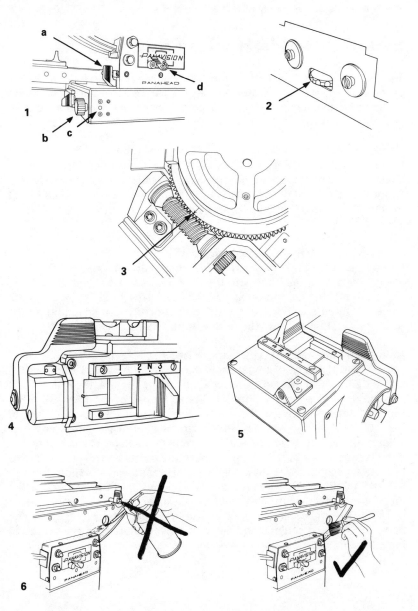

1. PANAHEAD adjustment controls: a. Pan feel (drag) adjustment/lock lever, b. Pan gear meshing adjustment and c. Release set screw, d. Tilt feel adjustment lever and quadrant lock, 2. Tilt tension adjustment wheel at front of PANAHEAD, 3. Index marks on pan gears which must be re-aligned if ever they become disengaged, 4. Pan gear ratio selector and pan lock, 5. Tilt gear ratio selector and tilt lock, 6. When a PANAHEAD has been in a dusty environment it is better to brush away grit rather than blow it farther into the gear teeth.

75

Special PANAHEAD Facilities

The skills of a Camera Operator may be greatly enhanced by making use of some of the additional features of the PANAHEAD in order to customize the head to the requirements of a particular shot without the need to exchange the head for another type.

Using the PANAHEAD as a free or friction head
For very fast "whip pans" the PANAHEAD may be used as a free head by attaching a pan bar to the rear of the head. An auxiliary pan bar may also be fitted to the front of the head. In this mode the gearing should be set to neutral. Resistance or "feel" may be introduced by tightening the "feel" adjusting levers.

Using the PANAHEAD as a gyro head
For long smooth pans the PANAHEAD may be used as a gyro head by using it as a free head but with the hand wheels fitted and the gearing set to the highest ratio to maximize the effect. The finger knob of the tilt wheel may be reversed to eliminate the possibility of it snagging on the operator's clothing. (Press the button at its center to release.)

Using the PANAHEAD as a remote head
For reasons of safety, or to get shots that would otherwise be impossible with a Camera Operator in close proximity to the camera, the PANAHEAD may be used as a remote-controlled head by use of the PANAREMOTE remote control system. In this mode slave motors are fitted in place of the PANAHEAD hand wheels and the camera controlled by remote hand-wheels, the aid of a Video Assist and remote lens and camera controls. The Operator may be 150 ft. away from the camera and still be able to operate the camera with all the subtlety of a hands-on situation.

The PANAREMOTE system also incorporates facilities to control the focus, zoom focal length and aperture of the lens and the camera on/off operation. All the controls, together with the video assist image, a video picture of the lens settings and film and other camera data are passed to the operating position along a single cable.

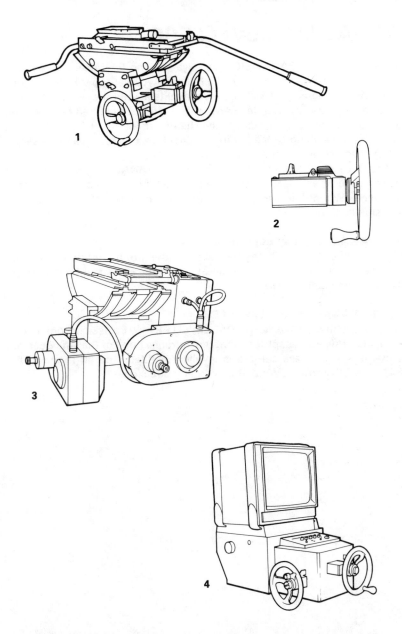

1. PANAHEAD with pan handles attached, 2. Tilt handwheel with knob turned round when PANAHEAD is in gyro-head mode, 3. PANAREMOTE remote pan and tilt system, remote control motors fitted in place of normal pan and tilt handles, 4. PANAREMOTE system, control console.

One person's level is another person's tilt.

PANABALL LEVELLERS

Yet another useful PANAVISION accessory is the PANABALL LEVELLER which eliminates the need to level a Mitchell-type flat-top tripod by short-ening or lengthening, or kicking, individual legs until the head is level.

For general use the regular PANABALL LEVELLER is fitted between the top of a Mitchell-type tripod top and the underside of a PANAHEAD and gives 15° of levelling capability in any direction. Equally it can be used to tilt the camera by 15° in any direction.

The three short feet on the underside of a PANABALL LEVELLER may be removed when it is fitted to an extra wide tripod, dolly or camera crane interface. (When doing so, unscrew the feet and screw them in from the top side so they do not get lost).

The PANABALL LEVELLER as a hi-hat
A PANABALL LEVELLER may be placed on the ground or fitted to any flat surface and used as a levelling hi-hat. The three short feet have holes in them for retaining screws.

The PANABALL Nodal LEVELLER
For SFX use the PANABALL NODAL LEVELLER may be used in conjunction with a PANANODE adaptor so that the entrance pupil of the lens does not move in space as the camera is levelled or tilted. The tilt range of the PANABALL NODAL LEVELLER is 3° in any direction.

78

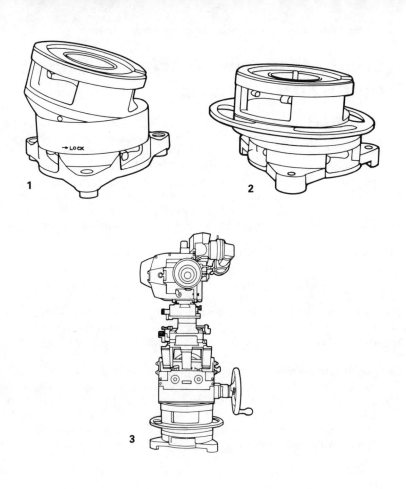

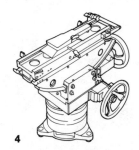

1. Regular (15°) PANABALL leveller, 2. Nodal (3°) PANABALL leveller, 3. Nodal PANABALL leveller/PANAHEAD/PANANODE/PANAFLEX assembly, 4. PANABALL leveller used as a hi-hat on the ground, 5. PANABALL foot reversed.

Nodally Mounting PANAFLEX and PANASTAR Cameras

Many aspects of the PANAFLEX CAMERA SYSTEM make it especially suitable for film productions with a major SFX content. Not least of these is where the lens is mounted relative to the pan and tilt movement.

For normal usage PANAFLEX and PANASTAR cameras are mounted centrally on a PANAHEAD and as low as is practical. However, for SFX work, especially when shooting miniatures and for Front Projection work, it is of crucial importance that the camera be mounted nodally on the head, that is, with the front entrance pupil of the lens (the point inside the lens where the bundle of light rays entering the lens appears to meet) is set exactly above the center of rotation of the pan axis and in the exact center of the tilt axis (the center of the quadrant).

To achieve this it is necessary to mount the camera about two inches higher than normal, about .4" to one side and to be able to slide the camera rearwards on the head until the entrance pupil of the lens (which is in a different place on every lens) is directly above the center of the pan axis, regardless of what this does to the balance and the "feel" of the head.

Setting the camera in a nodal position

The easy way to check if a camera and lens are mounted nodally is to place a post in front of the camera and note through the viewfinder its position relative to the studio wall or to the horizon. If the camera is correctly mounted it will be possible to pan and tilt the camera without changing the relative positions of the post and the mark on the studio wall.

80

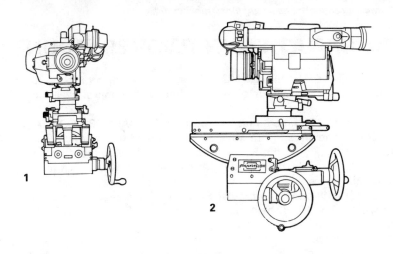

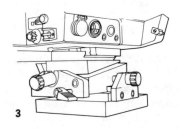

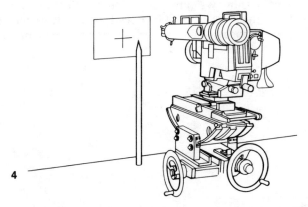

1. Camera mounted nodally above the pan axis and about the rotational axis,
2. Camera mounted nodally above the pan axis and about the tilt axis,
3. PANANODE micro-adjustment unit, 4. Setting a camera nodally by the use of two markers.

The mount can make this a very topsy turvy world.

The PANATATE TURNOVER MOUNT

The PANATATE TURNOVER MOUNT enables the camera to be turned sideways, upside down, or even rotated about the optical axis during a take.

The camera attachment unit incorporates a PANANODE NODAL ADAPTOR for fine height and sideways adjustment and the entire unit may be slid longitudinally on the PANAHEAD to make three-dimensional nodal positioning possible, making it an ideal instrument with which to photograph miniatures.

The unit is fitted with a regular PANAHEAD three-speed gearbox and handwheel. The handwheel may be interchanged with a PANAREMOTE actuator unit for remote control.

Camera Operators may find it easier to operate this unit with the aid of a PANAVID VIDEO ASSIST system, with a small monitor set on the unit, rather than try to look through the camera viewfinder as the camera is rotated.

The PANATATE system can also be used with a 65mm camera.

Fitting a camera to a PANATATE turnover mount
When fitting a PANAFLEX or PANASTAR camera to the PANATATE TURN-OVER MOUNT the film magazine must be fitted in the rear position.

The camera battery, control and video cables should be arranged so as to leave the unit from directly behind the camera and not allowed to wrap around the camera as it is rotated.

Balancing weights may be placed around the turnover ring to balance the camera radially, making it easier to turn the camera through 360° and to control it. Other weights may be fitted to the front of the unit to counterbalance the mass when it is moved rearwards on the head to nodally align the entrance pupil of the lens with the center of the pan and tilt axes.

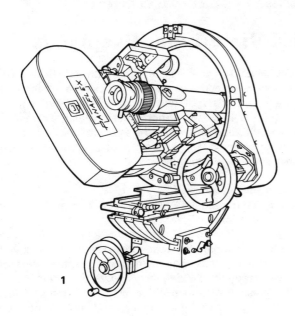

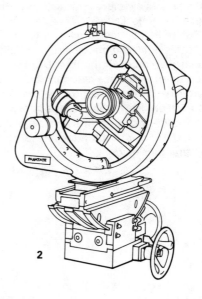

1. PANATATE unit, rear view showing turn-over operating handle, 2. PANATATE unit, front view showing counter weights.

What you see is what you get . . . especially with a PANAFLEX.

PANAFLEX Viewfinder Eyepieces

PANAFLEX (not PANAFLEX-X) cameras are fitted with INTERCHANGE-ABLE EYEPIECES. A short eyepiece is used when the camera is hand-held and supported on the operator's shoulder. A long-length viewfinder, in-corporating an IMAGE MAGNIFIER, is for normal use when the camera is mounted on a pan and tilt head and when the Operator needs to stand back from the camera when doing fast pans, etc., and an optional medium length which is somewhere between the two.

Fitting and interchanging eyepieces
To remove an eyepiece first unplug the PANACLEAR heated eyepiece and release the eyepiece leveller if fitted. Rotate the eyepiece until it is pointing directly upwards, tighten the friction lock at the front "elbow" of the view-finder system by turning it counterclockwise, press the thumb lock with your left thumb and turn the locking ring to the right until the reference lines coincide and then gently lift the eyepiece straight up to remove.

Fit an eyepiece by pressing it into position with the reference lines aligned and turn the locking ring to the left until an audible click indicates that the eyepiece is double locked in position. Do not over tighten.

Note: it is good practice to orient the viewfinder directly upwards when changing an eyepiece. This supports the weight of the extension eyepiece while it is unlocked.

Rotate friction lock clockwise to unlock.

At this point the image in the eyepiece will appear to be upside down. To correct, press in the thumb control below the horizontal arrow while rotating the eyepiece counter-clockwise until it faces downwards.

Release the thumb control, return the eyepiece (clockwise) to the upright position until an audible click is heard, and return to the shooting position.

Lock the friction ring if the eyepiece is to be self supported, leave it unlocked if it is to be attached to the eyepiece leveller or linked to the camera door hinge.

84

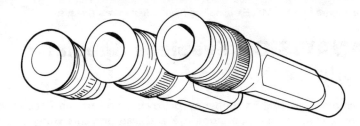

1. Set the three PANAFLEX interchangeable eyepieces.

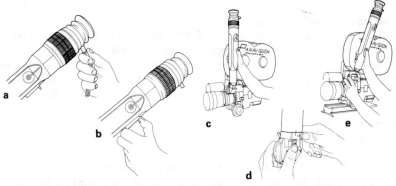

2. Removing a PANAFLEX eyepiece, stage by stage. . . .

a. Unplug the PANACLEAR, b. Release the eyepiece leveller, c. Set eyepiece
pointing directly up, d. Press in the thumb lock and turn locking ring to the right
until reference lines coincide, e. Remove eyepiece.

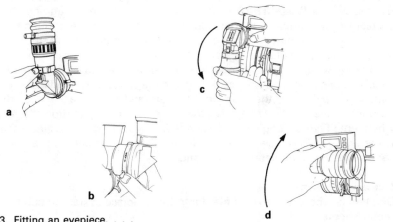

3. Fitting an eyepiece. . . .

a. Position eyepiece with reference marks aligned and turn locking ring to left,
b. Release friction lock, c. Press thumb control and rotate eyepiece clockwise
until pointing directly downwards, d. Release the thumb control and rotate the
eyepiece counter-clockwise, past the position where a click is heard, to the
shooting position.

85

PANAVISION Eyepiece Special Facilities

Ventilators in PANAFLEX rubber eyepiece cups prevent the discomfort of suction pulling on the operator's eye and also prevents water and dirt from entering the viewfinder optical system.

The ocular marker ring
PANAFLEX eyepieces have a WHITE MARKER BEZEL around their front rim so that the individual eyepiece settings of all the various people who must look through the camera may be marked with a pencil for easy future reference. This saves a great deal of time and petty aggravation when people with diverse eyesight are constantly using the viewfinder.

The image magnifier
The IMAGE MAGNIFIER enlarges the center of the ground glass image when it is necessary to carefully examine only a part of the scene. It may be used by the Director to check a particular performance during a rehearsal, by the Director of Photography to examine a lighting detail, by the Operator to study picture composition and especially by the Assistant when eye-focussing.

The PANACLEAR eyepiece de-mister
In cold weather conditions the PANAFLEX viewfinder optics may be heated by means of a PANACLEAR eyepiece heater to eliminate fogging. The problem of an eyepiece fogging during a take due to the comparative warmth of the operator's body or breath is a serious one and the PAN-ACLEAR heated viewfinder is a complete cure.

The PANACLEAR draws its power from the camera and is connected by a coiled cable plugged into a socket on the magazine port handle. As this handle is usually over the top magazine port (and the magazine is on the rear) when the short eyepiece is fitted for hand-holding and on the rear when a medium or long eyepiece is fitted, only a short connecting cable is necessary. A small switch and indicator light by the side of the outlet socket switches the heater unit ON and OFF.

Ocular diopters
Diopters can be supplied and fitted into the eyepiece system to suit user requirements.

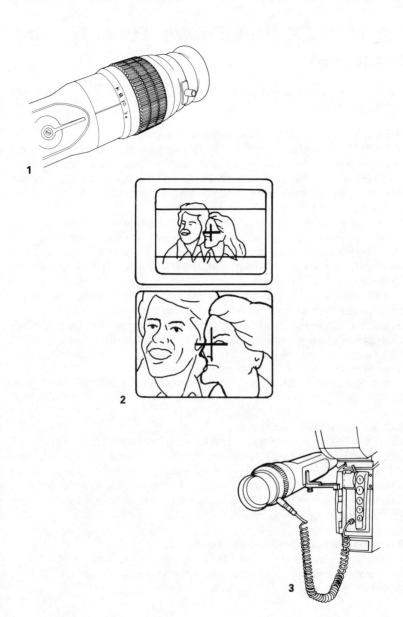

1. PANAFLEX eyepiece bezel showing multiple personal setting marks,
2. Simulated view through a PANAFLEX eyepiece showing normal and enlarged pictures using the magnification facility, 3. PANACLEAR power supply cable connected to an outlet on the magazine port handle.

PANAFLEX Viewfinder Tube Special Facilities

The Viewfinder Tube is that part of the viewfinder system which is attached to the camera.

The PANAGLOW illuminated reticle

PANAGLOW is PANAVISION's patented and unique illuminated reticle system which, when switched ON, causes the ground-glass markings to glow red so that the camera operator can clearly see the outline of the acceptable picture area, even in very low light conditions. It is an essential accessory when filming with low-key lighting.

The secret of the PANAGLOW system is its special ground-glass focusing screen, which uses discrete mirrors instead of etched lines to delineate the picture area. These mirrored lines reflect light to the operator from a small red lamp set in the viewfinder system. It is turned ON by a switch on the left side of the camera.

The intensity of the PANAGLOW light may be adjusted by a screw set behind a cover-screw on the front of the camera to the right of the engraved "PANAFLEX" nameplate. To alter the brightness, remove the protective cover-screw and use a small screwdriver to gently turn the adjusting screw. Take care not to set the brightness too high as this may cause the ground glass to "blush-out" with an overall red flare.

Later model PLATINUM PANAFLEXES have a knob to adjust the PANA-GLOW brightness level.

N.D. contrast viewing filters

All PANAFLEX cameras are fitted with two CONTRAST-VIEWING FILTERS which can be introduced into the viewfinder system to assist the Director of Photography in selecting the overall contrast and lighting balance of a scene. The filters normally supplied are ND 0.6 and 0.9. Alternative filters may be fitted on request. The Viewing-Filters are activated by a lever on the viewfinder tube. Press down for ND 0.6, up for ND 0.9.

De-anamorphoser and light cut-off

A lever on the side of the viewfinder tube may be set horizontal for spherical viewing, vertical for anamorphic viewing, or directly down for light cut-off.

88

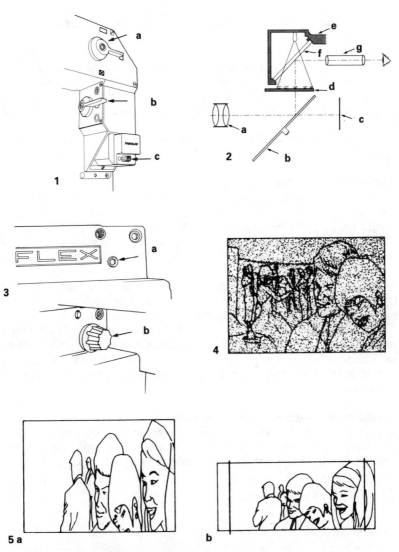

1. Front of PANAFLEX viewfinder system showing: a. the de-anamorphoser, normal and light cut-off, b. the viewing filter, and c. the PANAGLOW controls, 2. The way the PANAGLOW system works: a. Camera taking lens, b. mirror shutter, c. film, d. ground glass with mirrored markings which reflect light from the red LED lamp (e), f. partial mirror passes light from the LED to mirrored markings on the ground glass and at the same time reflects the ground-glass image to the viewfinder eyepiece (g), 3.a. Normal PANAGLOW brightness control and (b.) later model PLATINUM PANAFLEX type, 4. Effect of a viewfinder contrast filter, 5. Effect of the viewfinder de-anamorphoser: a. anamorphic image, b. unsqueezed image (note the 70mm frame lines).

89

PANAGLIDE Hand-Holding Facilities

All PANAFLEX and PANASTAR cameras, with the exception of the PANAFLEX-X, have been specifically designed from the very outset to afford the steadiest handholding possibilities and to sit on the Camera Operator's shoulder in an ergonomic manner with the center of gravity immediately above the spine to give a perfectly balanced feel and to minimize strain.

Alternatively, PANAFLEX cameras may be hand-held using a PANA-GLIDE floating camera system.

PANAGLIDE floating camera systems

The PANAGLIDE FLOATING CAMERA systems comprise a special suspension vest worn by the Camera Operator and either a special light-weight PANAFLEX camera or an even lighter weight Arri IIC camera.

Either system may be used inverted for a low camera position or may be attached to a moving vehicle for vibration-free photography from a bumpy ride.

PANAGLIDE vests are fitted with emergency releases so that the Camera Operator may divest himself of them in case of an emergency. This is especially important when working close to deep water.

Using a PANAGLIDE system successfully is a knack, just like riding a bicycle, and PANAVISION Inc. in Los Angeles and their Representatives worldwide are always happy to have any prospective PANAGLIDE operator visit their premises to try out a system, as long as there is a system available.

Hand-holding with maximum steadiness

The secret of hand-holding a camera with maximum steadiness is to use only wide-angle lenses. Other helpful tips are for the Operator to stand with his legs wide apart, to dig his elbows tight into his hips and to try to lean against a fixed object.

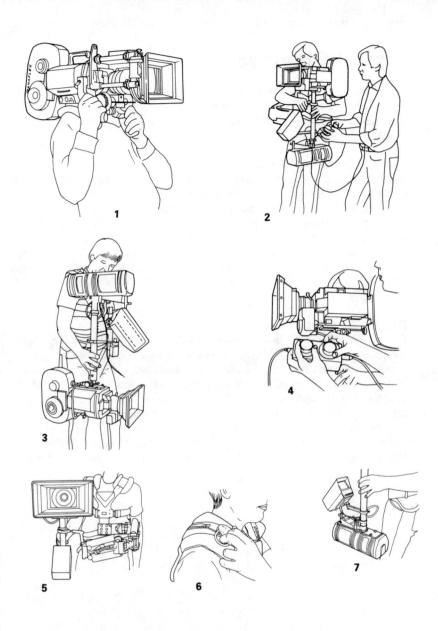

1. Hand-held PANAFLEX showing the elbows well tucked in for maximum steadiness, 2. PANAGLIDE in normal position, 3. PANAGLIDE in inverted position, 4. Special lightweight PANAFLEX-X and PANAVID, and remote lens focus and aperture controls used with the PANAGLIDE system, 5. PANAGLIDE camera support and harness, 6. PANAGLIDE emergency quick release, 7. PANAGLIDE counterweight/battery pack and monitor.

Digital Displays

Digital displays on all PANAFLEX and PANASTAR cameras indicate the actual running speed of the camera and function only when the camera is switched on and operating. Digital Footage Displays indicate the number of feet or meters of film shot since the counter was last reset.

PLATINUM PANAFLEX displays

The Digital Display on the PLATINUM PANAFLEX is double sided so that it can be viewed from either side of the camera. These two displays can be switched independently and may be set to show footage (feet or meters), camera speed, camera shutter angle, time code information and whether a behind-the-lens gelatin filter is fitted.

In addition the PLATINUM PANAFLEX and the PANAFLEX 16 cameras each have an ANNUNCIATOR PANEL, a row of LED warning lights which warn of LOW BATTERY VOLTAGE, INCORRECT CAMERA SPEED, FILM JAM, LOW FILM and OUT OF FILM.

PANAFLEX GII displays

On the PANAFLEX GII and earlier PANAFLEX and PANASTAR cameras the display is activated during operation and will remain lit for approximately five seconds after the camera is switched off. The footage shot display may be recalled by pressing the recall button on the top of the readout unit marked with a red "0". It may be reset by pressing the reset button at the left front end of the display, marked with a white "X," simultaneously with the recall button.

Between the displays are two LED lights. The lower, green lamp indicates that power is ON: the top, red light flashes when the battery voltage is 21-22 volts or less.

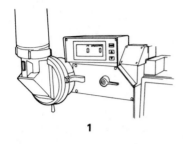 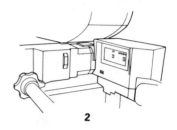

1. PLATINUM PANAFLEX digital displays and beam splitter cover plate,
2. Digital display on the reverse side of a PLATINUM PANAFLEX,

(a) Footage shots and fps

(b) Meters shot and fps

(c) Shutter angle and fps

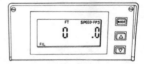

(d) Behind-the-lens filter in position

3. Alternative PLATINUM PANAFLEX displays

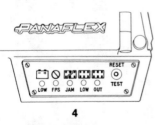

4. PLATINUM PANAFLEX annunciator panel (The PANAFLEX 16 is very similar),
5. PANAFLEX GII footage counter display.

93

The
Camera Assistants'
PANAFLEX

The Camera Assistants' PANAFLEX

PANAFLEX CAMERAS have been designed to simplify the mechanical side of the Camera Assistant's responsibilities and to allow him to concentrate on those aspects most directly concerned with creative film making.

The following section details the procedures for assembling, loading and operating PANAVISION PANAFLEX EQUIPMENT. We hope that even experienced PANAUSERS will discover a few valuable tricks and ideas.

Checking that the equipment is all there and is as required

Before starting on any operations to do with the camera, it is the Assistant's responsibility to check that the camera supplied by PANAVISION Inc., or by one of its representatives worldwide, as ordered by your Production company, is complete and is suitable to meet all the demands of the Production ahead. Remember that although the camera will have been meticulously checked out in the shop, the technicians carrying out that task may not have been informed of any special requirements.

Assistants are advised to ask their Production Manager beforehand for a list of the equipment to be made available to them and to report any changes, additions or deletions that are made during the testing period.

The Assistant should also look to see if the camera is fitted with an aperture matte, and if so that it is the correct one, and that a correctly marked ground glass is fitted.

Checklist

PANAVISION technicians take particular pride in preparing cameras for use by clients and are required to sign a special checklist form to confirm that the flange focal depth setting and the collimation to the ground glass are both within $1/10,000''$ of standard, that the camera runs free of scratches, that it has been tested for quiet running in the sound test room, that the electronics, the mechanics and the video are all operating as normal and that the film transport movement is clean.

They must even sign that the camera is "cosmetically" up to PANAVISION's high standard. It is all part of the pride of PANAVISION.

This card is included with the camera outfit when it leaves the shop and on the reverse the user is asked to state if all the equipment was delivered on time, if it was in a satisfactory condition and if the user has any comments to make.

PANAVISION particularly asks users to tell them if they detect the slightest thing going wrong with their equipment, as this enables them to make sure it is put right before it goes to another user, it being much easier to detect the beginnings of any malfunction when it is actually in use. Users are then asked to return the card to the Camera Service Department. PANAVISION is always grateful for user feedback.

96

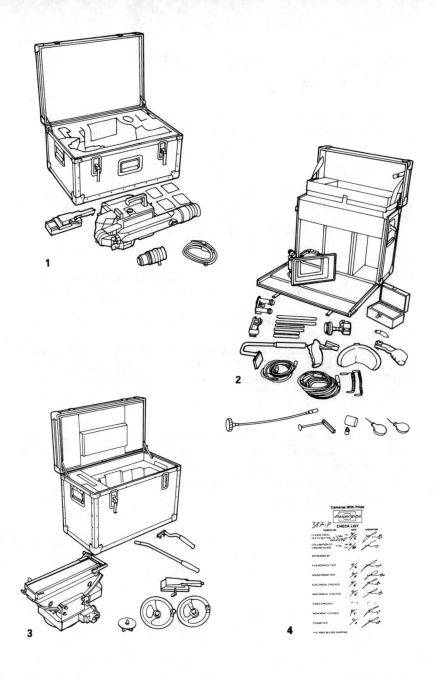

1. The contents of a PANAFLEX camera case, 2. The contents of a PANAFLEX camera accessory case, 3. The contents of a PANAHEAD case, 4. Checklist.

97

Setting up a PANAVISION PANAHEAD

The PANAVISION PANAHEAD geared head fits on any regular tripod, dolly or camera crane equipped with a standard Mitchell-type interface.

When setting the PANAHEAD in place be sure that the male locating key on the underside of the head is properly located in the keyway of the tripod and secure it by using the locking knob supplied.

Attach the pan and tilt handwheels by the central locking knobs after aligning the keyways correctly. The handwheel with the removable finger knob should be fitted in the tilt (rear) position.

Release the pan and tilt locking levers and check for smoothness of movement in all three speed ratios. In the gyro head mode the handwheels act as flywheels to smooth out the pan and tilt movements. The finger knob of the tilt handwheel may be reversed by pressing its center section and fitting it in the rear side of the flywheel so that it does not snag on the operator.

A two-way level is fitted to the rear of the PANAHEAD to check that it is set level.

A PANABALL leveller may be used between the head and a tripod to make it easier to level the head or to shoot with a tilted camera. A PAN-ABALL leveller may also be used as a hi-hat for fitting a head to a flat surface.

Fitting the tilt plate

Additional up or down tilt movements are possible by adjusting the tilt-plate system, which may be located either way. It is normally fitted with the pan-handle socket to the rear, giving additional downwards tilt. It may be reversed, with the pan-handle socket located at the front, to give additional tilt up.

98

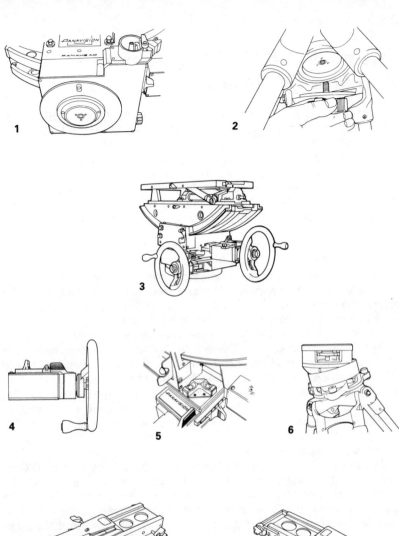

1. Underside view of a PANAHEAD showing the male locating key, 2. Head secured by means of a central locking knob, 3. (Super) PANAHEAD handwheels, 4. Reversed tilt handwheel knob, 5. Two-way level, 6. PANABALL leveller, 7. Tiltplate with pan handle socket at the rear (normal) for additional tilt down, 8. Tiltplate with pan handle socket at the front for additional tilt up.

Fitting a PANAFLEX Camera to a Pan and Tilt Head

BEFORE FITTING A CAMERA TO A SLIDING BASE PLATE FIRST MAKE SURE THAT THE BASE PLATE IS SECURELY LOCKED IN ITS SLIDE.

To attach a PANAFLEX camera body to a PANAHEAD or a Sliding Base Plate make sure that the tail lock is completely withdrawn by fully turning it counter-clockwise and downwards; place the camera body squarely on the front of the dovetail-shaped interface and push forwards until an audible click is heard, indicating the camera is firmly in position; rotate the tail lock knob clockwise and tighten until the rubber pad is firmly against the rear of the camera.

Earlier PANAHEADS and Sliding Base Plates have a tail lock which must be unscrewed to release rather than turned to one side.

When the camera has been completely assembled with a loaded magazine, a lens and other appropriate accessories, the entire unit may be slid forwards and backwards to bring it into balance by releasing the balancing slide lock lever on the left side of the dovetail plate.

Securely lock off the balancing slide when proper balance is achieved.

To remove a PANAFLEX camera from a PANAHEAD or Sliding Base Plate, position the viewfinder fully forward (resting on the matte box), hold the camera body firmly by the handle on top of the camera, release and turn the tail lock counter-clockwise until it is clear of the camera, pull the release catch at the side of the dovetail, move the camera back and lift carefully off the base, using both hands to do so.

Using a sliding base plate

PANAFLEX cameras may be fitted to any other flat top tripod head (O'Connor, Ronford, Sachtler, Vinten, etc.) by the use of an optional Sliding Base Plate assembly which provides an interface between a standard ⅜" 16 TPI tripod screw and the PANAVISION Dovetail attachment system. Note: A Sliding Base Plate is always supplied and is to be found in the camera case.

100

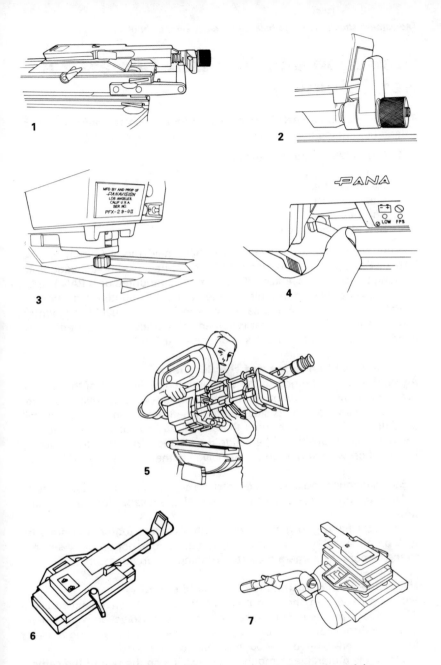

1. Top of PANAHEAD with tail lock set to one side, 2. Tail lock set tight, 3. Earlier type tail lock, 4. Side release lever, 5. Removing camera from a PANAHEAD using two hands, 6. Sliding base plate, 7. Sliding base plate on fluid head.

101

Preparing a PANAFLEX Camera for Hand-Holding

Before converting a PANAFLEX camera from a mounted configuration to a hand-held mode it is necessary to have the following items at hand:

1. 500' or 250' magazine (usually)
2. Right hand grip
3. Left hand grip
4. Shoulder rest
5. Short eyepiece
6. Belt or on-board battery.

If the camera is to be used in the hand-held mode (except for the PANAFLEX-X, which is not intended for hand-held use) first change from the long to the short eyepiece. If the camera is mounted on a PANAHEAD it is usually easier to make this change before dismounting the camera.

Similarly, it may be found easier to change from a 1000' top-mounted magazine to a 250' or 500' rear mounted configuration and to fit the right hand grip while the camera is still on a PANAHEAD.

Attaching the handgrips and shoulder rest

Attach the right hand grip to the slide situated at the front of the camera to the left of the lens mount, turning the locking lever clockwise to secure.

Turn the camera onto its side and fit the left hand grip into the dovetail slide (located underneath) until there is an audible click, indicating it is safely locked in position. Then slide the shoulder rest into the slideway located crosswise at the top of the "cave" in the middle underside of the camera.

Both handgrips and the shoulder rest are adjustable for operator comfort.

An optional extension for added control is available for the left hand grip.

The right hand grip is fitted with both a camera toggle and pressure type on-off switches. Note: If the right hand grip is not properly secured to the camera the camera may stop running or run intermittently.

Magazine considerations for hand-held shooting

For most hand-held shooting it will be found most convenient to use a 500' magazine mounted on the rear of the camera. However, when weight is an important consideration a 250' magazine can be used, and if a very long take is envisioned, a 1000' magazine can be fitted.

While it is preferable to mount the magazine on the rear of the camera for optimum balance, it may be mounted on top if it is necessary for the operator to get as far back as possible.

102

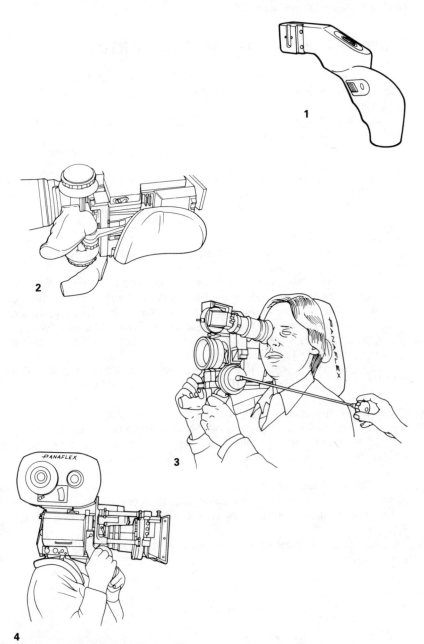

1. Right-hand handgrip, showing switches, 2. Underside of camera showing hand grips and shoulder rest fitted, 3. Hand-held camera with a rear mounted 500′ magazine and with a follow-focus extension fitted, 4. Hand-held camera with magazine on top.

Little and often is the best way.

Camera Movement Lubrication

In general, camera movements should be lightly lubricated at the start of each working day using *one drop* of the special oil supplied by PANAVISION in each oil well.

If more than 20,000' of film (6000m) is shot per day then it is advisable to oil the camera more often.

When cameras are run at speeds greater than 24 fps they should be lubricated more often.

If, when lubricating a camera, a regular clockwise progression is made, there is less chance that any oil well will be missed.

IMPORTANT: DO NOT *OVER-LUBRICATE* CAMERA MOVEMENTS.

PANAFLEX lubrication points
There are 12 lubrication points on all 35mm PANAFLEX movements irrespective of model type.

In addition, the thin metal strip between the bottoms of the pull-down claw slots in the register plate should be lightly pressed to check that the felt pads at the bottom of each slot are slightly moist. These pads are designed to smear a minute amount of silicone liquid onto the undersides of the pull down claws and if dry should be carefully moistened using a drop of PANAVISION *silicone* liquid.

IT IS VERY IMPORTANT NOT TO SPILL ANY SILICONE LIQUID ONTO THE CAMERA MOVEMENT AS THIS MAY CAUSE DAMAGE.

PANASTAR lubrication points
There are eight lubrication points on PANASTAR camera movements.

When the camera is run faster than 24 fps it should be oiled more often, and at maximum speed should be oiled before every take.

PANAFLEX 16 lubrication points
There are 12 lubrication points on PANAFLEX 16 cameras.

To lubricate the eccentric pivot arm it is necessary to remove the aperture plate.

To lubricate the lower register pin bush it is necessary to remove the pressure plate.

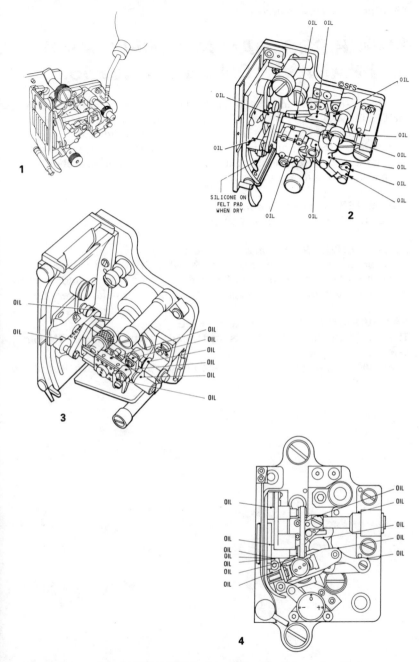

1. Use only one drop of PANAVISION oil in each oil well from the dispenser provided with each camera, 2. PANAFLEX lubrication points, 3. PANASTAR lubrication points, 4. PANAFLEX 16 lubrication points.

105

These also must be oiled every day.

SUPER PSR, PAN-MITCHELL Mk.II and PANAVISION 65 Lubrication

Like the PANAFLEX and PANASTAR cameras, the SUPER PSR, PAN-MITCHELL MK.II and PANAVISION 65mm cameras should be lubricated daily in a routine manner so that no oil wells are overlooked.

In addition, the pair of skew gears on the main drive shaft on the motor side of these cameras should be lubricated with a light grease about once per month with average use.

SUPER PSR lubrication points
There are 13 lubrication points on all types of PSR cameras.

PAN-MITCHELL Mk.II lubrication points
There are seven lubrication points on PAN-MITCHELL Mk.II cameras.

When these cameras are run faster than 24 fps they should be oiled more often, and at maximum speed should be oiled before every take.

PANAVISION 65mm lubrication points
There are eight lubrication points on a Standard type PANAVISION 65mm camera.

In addition a single drop of oil should be applied to each end of the pressure plate rollers.

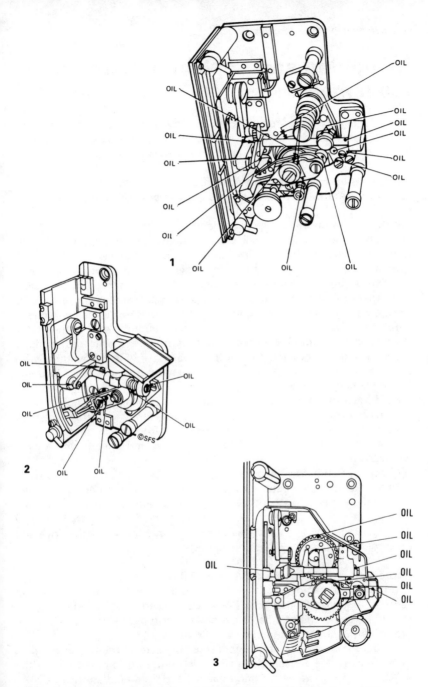

1. SUPER PSR lubrication points, 2. PAN-MITCHELL Mk.II lubrication points,
3. PANAVISION 65mm lubrication points.

107

Removing and Replacing Camera Movements

Movements may be removed from the camera body for lubricating and cleaning, provided that the flange focal depth is checked before use. Otherwise movements should be cleaned and oiled while still in the camera.

Removing and replacing a PANAFLEX camera movement

Tilt the camera slightly up to ensure the camera door remains open on its own.

Inch camera until the pulldown claws are at the bottom of the stroke and fully withdrawn from their slots, stopping just before the pulldown claw arm obstructs the lower movement lockdown screw.

Gently use a wide screwdriver to loosen the two short knurled-head "captive" screws which secure the movement plate. Unscrew approximately five turns until they go loose in their bushings. Remove movement by pulling on the pitch control knob, wiggling to loosen. Please be gentle.

To replace the movement, inch the camera until the pins of the motor coupling are horizontal with the witness mark downwards and similarly align the movement shaft so that they match. Hold the movement with both hands using the left hand for support and the right hand to guide the movement into position.

Hold the movement with the thumb on the pitch control knob and the forefinger on the top aperture dog-lock, slide the entire unit into the camera interior, engage coupling in camera body with the witness marks aligned, secure with short knurled-head capture screws and tighten with a wide screwdriver.

Check the flange focal depth (see page 150) before reusing the camera.

Note: The interface between the motor drive coupling and the movement coupling is off-set so they cannot be assembled incorrectly. If the movement does not seat, inch camera back and forth slightly until they fit snugly together. If it still does not seat remove and check the drive and movement couplings for possible damage.

Removing and replacing a PANASTAR camera movement

The method of removing and replacing the movement of a PANASTAR camera is similar to that of a PANAFLEX except that pull down claws should be set in the middle of the pull-down movement and then withdrawn by pushing the retraction knob downwards.

In addition to the two movement retaining screws there is also a cam at the rear of the movement plate which must be released before the movement can be removed and replaced and tightened afterwards.

108

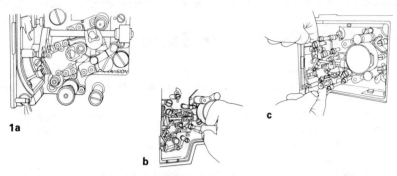

1a

b

c

1. Removing a PANAFLEX movement: a. Position the claws at the bottom of the pull-down stroke stopping just before the link arm covers the lower captive screw, b. Loosen the two captive screws with a broad screwdriver, c. Pull gently on the pitch control to remove the movement.

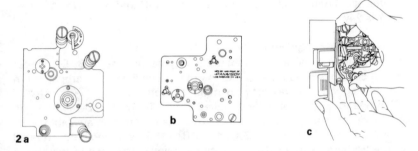

2 a

b

c

2. Replacing a PANAFLEX movement: a. Inch camera until witness mark on the movement coupling is directly downwards, b. Align the movement until the witness mark is directly downwards, c. Hold movement correctly to replace.

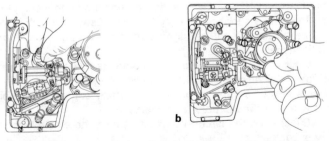

3 a

b

3. Removing a PANASTAR movement: a. Pull down claws should be in the middle of the pull-down movement and then withdrawn by pushing the retraction knob downwards, b. In addition to loosening the captured screws a cam at the rear of the movement plate must also be released with a screwdriver.

Cameras need to be fed with electricity as well as filmstock.

PANAFLEX Power Supplies

All PANAVISION cameras, except the PSR, operate from a 24 volt DC power source which may be a 24 volt battery or a PANAVISION 100-120/200-240 volt 60-50 Hz Battery Eliminator.

All PANAVISION camera-to-battery cables use the same 3-pin connector at the battery end of the cable (pin 1 is 24v+, pin 2 is ground) although the connector at the camera end may differ depending upon the camera and the model.

On PANAFLEX 16 (and on some PLATINUM PANAFLEX) cameras the power supply is attached to the camera via a 2-pin LEMO connector, the right hand one of three sockets situated at the bottom left hand corner of the right hand side of the camera. (Pin 1 is 24v+, pin 2 is ground.)

On other PANAFLEX and PSR cameras the power supply is attached to the camera via a 3-pin LEMO connector, the center one of three sockets situated at the bottom left hand corner of the right hand side of the camera. (Pin 1 is 24v+, pin 2 is ground.)

Camera power requirements

The minimum power requirement is 21v DC, the maximum 28v DC. Under normal circumstances most cameras draw approximately 2 amps, but in very cold conditions may draw as much as 9 amps.

The battery packs normally supplied are 10-16 Ah. Light-weight 4 and 6 Ah shoulder-slung, belt and on-board batteries are also available.

On the PLATINUM PANAFLEX a LOW BATTERY warning light, situated on the left hand side of the annunciator panel, shows when the power supply is low or inadequate (21 volts or less).

On the GII/GOLDEN PANAFLEX a green battery condition light, adjacent to the footage/fps display shows when the power supply is adequate; a red light indicates that power is low (21 volts or less).

Camera power notes

The camera must NEVER be switched on or off by connecting or disconnecting the power supply. Use only the power switch which is located at the rear of the camera, the handgrip switch or the optional side or remote control switches.

The camera ON-OFF switch only disconnects power to the camera motor. It does NOT disconnect battery power to the camera.

There is always a small drain on the battery caused by the electronic circuits whenever the battery is connected.

DO NOT change the circuit boards or touch the electronics while the battery is connected.

Disconnect the camera from the battery during long breaks.

110

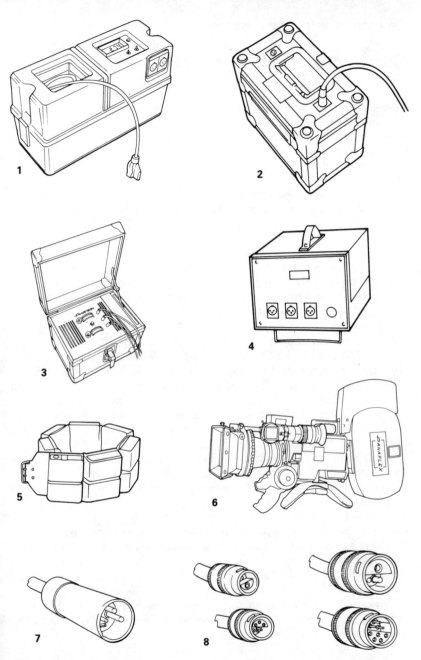

1. Regular LA ''purse'' battery complete with charger, 2. Regular Nicad battery,
3. Battery charger for Nicad batteries, 4. Battery eliminator, 5. Battery belt,
6. On-board battery, 7. Universal 3-pin battery connector, 8. Various LEMO
connectors.

111

Recharging Nicad Batteries

PANAVISION cameras may be supplied either with Nickel Cadmium or Sealed Lead Acid batteries. Both types have a good power-to-weight ratio, are maintenance free and hold their charge well under normal temperature conditions. From the users' point of view the principal differences are in the voltage of the individual cells, the type of charger that needs to be used and the effect of habitual overcharging.

Nickel cadmium batteries

PANAVISION supplies two types of Nicad Batteries, the normal type being made up of cylindrical cells and the higher capacity type which uses rectangular nickel-cadmium alkaline cells.

Cylindrical cells, being smaller, are most usually used for battery belts and for on-board batteries.

All Nicad batteries hold their charge best when kept cool and deliver best when warm. They self discharge quite rapidly at temperatures above 95°F (35°C) but below 32° (0°C) can hold their charge for years.

They give of their maximum at 85-95°F (30-35°C). At 32°F (0°C) they give only half capacity and at −40° may not operate at all.

The nominal voltage of Nicad cells is 1.2v. When charging is just completed, for a short period the voltage may be as high as 1.5v per cell. For this reason they should be rested for at least two hours after charging to allow the voltage to stabilize.

On discharge Nicad cells maintain a very constant 1.2v per cell until they come to the end of their charge when the voltage drops rapidly. They should not be discharged below 1v per cell.

Nicad batteries must only be recharged using a constant-current type charger designed especially for recharging Nicad type cells.

If Nicad batteries are frequently recharged before they have been fully discharged they will suffer from what is known as "memory effect" and will cease to hold their full capacity. This state can be corrected by discharging the batteries to 1v per cell several times between full charging. It is very bad practice to recharge Nicad batteries when they are already fully charged.

PANAVISION Inc. in Tarzana, California, has a special Nicad battery testing rig which plots the current flow of batteries as they are fully discharged under controlled conditions. On this device any malfunction, even of a single cell, is plotted on a graph so that any faulty cell may be identified and replaced.

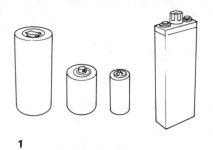

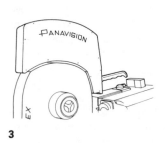

1

2

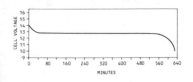

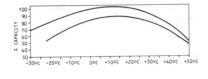

3

4

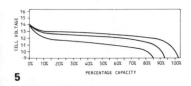

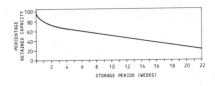

5

1. Cylindrical and rectangular Nicad battery cells, 2. Battery belt made up of Nicad cells, 3. On-board battery made up of Nicad cells, 4. Nicad battery test rig, 5. Nicad battery graphs.

113

Recharging Sealed Lead Acid Batteries

The sealed lead acid (LA) batteries supplied by PANAVISION for the purpose of powering cameras and camera accessories are far removed from the lead acid type batteries used for automobiles. Unlike the common lead acid battery the sulfuric acid electrolyte of LA batteries is solid and cannot be spilled. From the outside they look very similar to cylindrical type Nicad batteries but that is where the similarity ends.

The nominal voltage of LA cells is 2 volts. Twelve LA cells are required to make up a 24v battery compared to twenty Nicad cells.

LA batteries must be charged at approximately 2.5v per cell using a constant-voltage charger. For this reason constant-current chargers designed to be used with Nicad batteries are not suitable for use with LA batteries, and vice versa. Only chargers supplied by PANAVISION for use with particular types of batteries should be used with those batteries. Many batteries supplied by PANAVISION incorporate chargers within the battery case so there can be no confusion.

Charge retention and performance

Like Nicad batteries, LA batteries store their charge best when kept in a cool atmosphere but unlike Nicads should only be stored in a fully charged state. They will, however, store their charge better than Nicads in warm conditions and will continue to perform at temperatures slightly lower than Nicads.

Unlike Nicad batteries, LA batteries do not suffer from the memory effect. Furthermore, under-discharging between recharging will increase the number of charge-discharge cycles to be expected from the life of a particular battery.

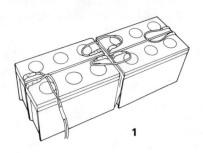

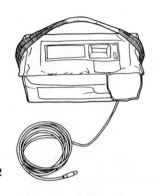

SEALED LEAD ACID BATTERY TEMPERATURE v CAPACITY TABLE

Temperature		Capacity
°C	°F	
61	140	110%
38	100	105%
20	68	100%
0	32	85%
−12	10	50%
−40	−40	15%

1. Connected blocks of LA cells, 2. LA battery with built-in charger, 3. The voltage/amperage meter of an LA battery.

115

Few things are more important than the correct camera speed.

Setting the Camera Speed

Before running the camera, especially for the first time during the day, it is a wise precaution to check that the camera speed (fps) is set as required.

The 24/25, 24/29.97 or 24/30 crystal control fps switch is situated in the recess in the center underside of the camera (the cave). On most cameras this switch may be configured to switch between either 24 and 25 fps or 24 and 29.97/30 fps by means of a switch on the crystal/tach circuit board. The latest circuit boards allow you to switch between all four speeds.

PLATINUM PANAFLEX speed controls

The PLATINUM can be set to run under crystal control at any speed (4-36 fps) in increments of $\frac{1}{10}$ fps. by a switch at the rear of the camera. (The 29.97 fps switch is in the cave.) Lift the cover to set the speed and check that it is exactly as required by running the camera and observing the digital display panel. (Note: At 29.97 fps the display will show 29.9 and flash.)

To vary the camera speed continuously and smoothly over the full range a special Variable Speed Control Unit is an available accessory.

GII and earlier PANAFLEX speed controls

The crystal control/variable speed switch of these cameras is located at the rear of the camera providing 4-36 fps variable speeds. Push the slide switch to the "VARY" position (a safety catch prevents this from being done inadvertently) and turn the knurled black knob to vary the speed. On some cameras the maximum speed may be achieved by turning the knob fully clockwise and then backing off very slightly. Running speed is confirmed by the LED display.

PANASTAR speed controls

A switch at the rear of PANASTAR cameras may be used to set the camera to operate at crystal controlled speeds of 6, 12, 18, 24, 36, 48, 60, 72, 96 and 120 fps.

The 24 fps selection may be set to run at 25 fps by means of a switch in the cave and on some cameras this setting may in turn be set to operate at 30 fps by a switch on the crystal/tach board.

Other pre-set crystal controlled speeds and infinitely variable speeds may be selected by means of an external speed control unit.

On PLATINUM PANASTAR cameras a digital switch may be used to set the camera to run at any crystal controlled speed between 4 and 120 fps.

Precision speed control

An electronic Precision Speed Control is available to run any camera at any speed, in increments of $\frac{1}{1000}$ fps, under crystal control.

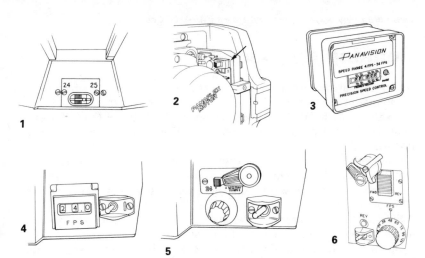

1. 24/25, 24/29.97 or 24/30 fps switch set in camera cave, 2. 25/30 or 25/29.97/30 fps configuration switch on the circuit board, 3. Precision Speed Control accessory, 4. PLATINUM digital speed selector and rear ON/OFF switch, 5. GII/GOLDEN crystal control (24, 25, 29.97 or 30 fps)/variable speed switch, variable speed control and rear ON/OFF switch, 6. PANASTAR speed control.

HMI FLICKER-FREE CAMERA SPEEDS

	LIGHT PEAKS	SHUTTER ANGLE										
	PER	40	60	80	100	120	140	144	160	172.8	180	200
Hz	EXPOSURE	FRAMES PER SECOND (To the nearest 0.1 fps)										
50	1*	22.2	33.3	44.4	55.6	66.7	77.8	80	88.9	96	100	111.1
50	1	11.1	16.7	22.2	27.8	33.3	38.9	40	44.4	48	50	55.6
50	1.5	7.4	11.1	14.8	18.5	22.2	25.9	26.7	29.6	32	33.3	37
50	2	5.6	8.3	11.1	13.9	16.7	19.4	20	22.2	24	25	27.8
50	2.5	4.4	6.7	8.9	11.1	13.3	15.6	16	17.8	19.2	20	22.2
50	3	3.7	5.6	7.4	9.3	11.1	13	13.3	14.8	16	16.7	18.5
50	4	2.8	4.2	5.6	6.9	8.3	9.7	10	11.1	12	12.5	13.9
60	1*	26.7	40	53.3	66.7	80	93.3	96	106.7	115.2	120	133.3
60	1	13.3	20	26.7	33.3	40	46.7	48	53.3	57.6	60	66.7
60	1.5	8.9	13.3	17.8	22.2	26.7	31.1	32	35.6	38.4	40	44.4
60	2	6.7	10	13.3	16.7	20	23.3	24	26.7	28.8	30	33.3
60	2.5	5.3	8	10.7	13.3	16	18.7	19.2	21.3	23	24	26.7
60	3	4.4	6.7	8.9	11.1	13.3	15.6	16	17.8	19.2	20	22.2
60	4	3.3	5	6.7	8.3	10	11.7	12	13.3	14.4	15	16.7

Note: The above speed settings are dependent upon very accurate frequency and shutter settings. Where a speed is rounded off it is advisable to use a micro shutter control and a look-through device to ensure safe flicker-free operation (see page 121).

A 200° (or any other) shutter opening is safe at 24 fps with a 60 Hz AC power supply providing that both the camera speed and the 60Hz power supply are very precise.

*One light peak during the exposure period but none during the viewfinder period.

117

There are many ways to start and stop a PANAFLEX camera.

Running the Camera

Under normal circumstances the camera is turned ON and OFF by a switch at the rear of the camera.

On the PLATINUM PANAFLEX an optional additional ON/OFF switch may be situated on the viewfinder side of the camera just below the PANAGLOW switch.

An extension power switch for remote control is an optional accessory and slides into the dovetail situated to the side of the lens port.

Certain of the optional electronic accessories also provide camera ON/OFF capabilities.

Slack film in the rear side of the magazine may be taken up by pressing in the inching knob on the rear of the camera.

On the PLATINUM, additional inching and run switches are situated inside the camera film compartment, just below the main sprocket, to make it easier for the assistant to check that the film is threaded properly and the camera is running correctly before closing the camera door.

Running the camera at the beginning of the day

Prior to each day's shooting, run the camera without film for a short period to ensure that the mechanism is functioning smoothly. If the camera is fitted with a micro switch situated at the bottom right-hand corner of the film compartment opening this must be depressed to run the camera with the door open.

Running a PLATINUM PANAFLEX or a PANASTAR in reverse

PLATINUM PANAFLEX and some PANASTAR cameras may be run in reverse using special reverse running magazines. It should be noted that the film winds on or off the rear roll of film in the magazine in the opposite direction to normal.

Reverse running may be used either for shooting a scene to run backwards on the screen (in which case there must be sufficient unexposed film loaded in the rear of the magazine), for rewinding the film between normal direction multiple exposures or for some two-camera 3-D setups.

To run the camera in reverse a special switch to the left of the rear ON/OFF switch must be set to the REV position.

118

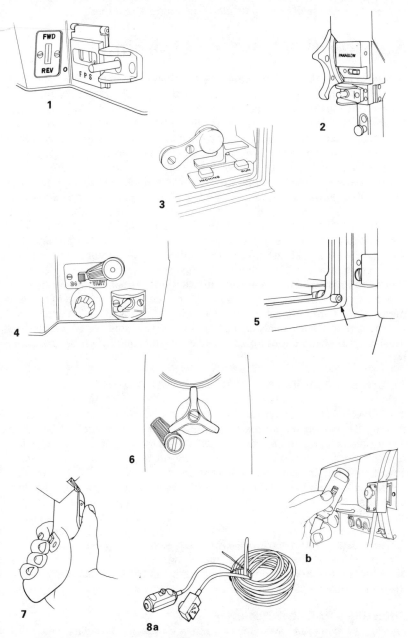

1. PLATINUM reverse running, digital speed selector and rear ON/OFF switch,
2. PLATINUM optional side ON-OFF switch, 3. PLATINUM internal INCHING and
RUN switches, 4. GII ON-OFF switch, 5. Micro switch at bottom right corner of
camera body (GII and earlier cameras), 6. Inching knob, 7. Camera handle
switches, 8 a. & b. Camera extension switch.

119

Setting the Shutter Opening

A feature of all PANAFLEX cameras is that they have separate focal plane and mirror shutters. The advantages of this configuration are that the exposure time can be adjusted, in shot if necessary, by adjusting the shutter opening to any angle between 50° and 200°, giving up to two stops of exposure control and allowing the exposure period to be set accurately to suit HMI lighting and synchronization with video and computer displays.

On PANASTAR cameras the adjustable shutter range is 40-180°.

The opening of the focal plane shutters of all PANAFLEX cameras may be adjusted by use of a quadrant control at the top rear of the camera to the right of the rear magazine port. For coarse adjustment the lever may be set according to the engraved markings and locked off as required.

Adjustable limit stops at either end of the quadrant make possible in-shot exposure adjustment from one shutter opening to another without the need to look at the quadrant markings while so doing.

For the most critical adjustment of the shutter opening, as when synchronizing the exposure period to exactly match the scan period of a video monitor or a computer, an Aperture Viewing Mirror is available which can see through the lens exactly as the film does. To use this device the camera must be run at the shooting speed with the film and the pressure plate removed.

Adjusting a PLATINUM PANAFLEX or PANASTAR shutter
Before adjusting the variable shutter on a PLATINUM PANAFLEX or PLATINUM PANASTAR it is first necessary to open the camera door and release the shutter locking lever situated just above the movement.

For precise adjustment press the MODE button on the Digital Indicator repeatedly until it displays the SHUTTER ANGLE. Run the camera and adjust the shutter using the quadrant control at the rear of the camera. When the digital display indicates the precise shutter angle required lock off the shutter setting by means of the shutter locking lever inside the camera.

Adjusting a GII and earlier PANAFLEX shutter
On the GII and earlier cameras a micro adjustment device which screws onto the shutter adjustment quadrant is available for precise setting of the shutter opening.

Adjusting a PANASTAR shutter
PANASTAR cameras have a locking screw on the shutter adjustment control on the rear of the camera. To set the shutter, release the locking screw, set the shutter as required and re-lock.

Certain frequently used precise shutter openings are engraved on the shutter blades and may be seen by removing the film and pressure plate.

120

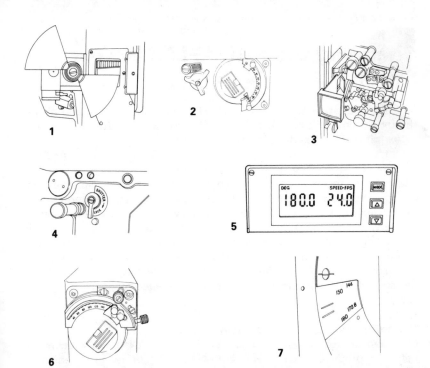

1. PANAFLEX with front removed showing focal plane and mirror shutters,
2. PANAFLEX shutter control quadrant on rear of camera, 3. Aperture Viewing
Mirror in position, 4. PLATINUM PANAFLEX shutter release lever, 5. PLATINUM
PANAFLEX digit shutter-angle display, 6. Shutter quadrant micro adjuster,
7. Shutter openings engraved on the shutter blade of a PANASTAR camera.

HMI FLICKER-FREE OPTIMUM SHUTTER OPENINGS

Hz	LIGHT PEAKS PER EXPOSURE	FRAMES PER SECOND											
		16	18	20	22	24	25	30	40	50	60	100	120
		SHUTTER ANGLE											
50	1*	28.8	32.4	36	39.6	43.2	45	54	72	90	108	180	
50	1	57.6	64.8	72	79.2	86.4	90	108	144	180			
50	1.5	86.4	97.2	108	118.8	129.6	135	162					
50	2	115.2	129.6	144	158.4	172.8	180						
50	2.5	144	162	180	198								
50	3	172.8	194.4										
60	1*	24	27	30	33	36	37.5	45	60	75	90	150	180
60	1	48	54	60	66	72	75	90	120	150	180		
60	1.5	72	81	90	99	108	112.5	135	180				
60	2	96	108	120	132	144	150	180					
60	2.5	120	135	150	165	180	187.5						
60	3	144	162	180	198								
60	4	192											

*One light peak during the exposure period but none during the viewfinder period.

121

Changing the PANAFLEX Reflex System

An additional advantage of PANAVISION's dual focal plane and mirror shutter system is the possibility it affords to exchange the normal rotating reflex mirror for a fixed-pellicle reflex mirror.

Although now superseded by the PANAVISION flicker-free PANAVID CCD video assist system, PANAFLEX and PANASTAR cameras can achieve a greatly enhanced, flicker-free video assist image by replacing the spinning mirror, which reflects light to the reflex viewing system intermittently for only 44% of the exposure/pulldown time with a membrane-thin partial mirror which reflects approximately 33% of the light all of the time. It is particularly advantageous when using a PANAGLIDE floating camera system when an optimum quality video assist image is very important.

To compensate for the light that is diverted to the viewfinder by the pellicle reflex system the lens stop must be opened by approximately ⅓ stop.

Changing and handling a pellicle mirror

Changing the reflex system must be done by a PANAVISION technician before a production commences. PANAVISION or its representatives worldwide will give instructions on the use and handling of pellicle reflex mirrors whenever one is requested.

Camera Assistants must be most careful when using a pellicle reflex to ensure that it remains perfectly dust free (a blob of dirt will cast a shadow on the film) and that they do not poke their finger through it when changing the ground glass or for any other reason. The camera should be handled with extra care lest the pellicle mirror be shattered.

122

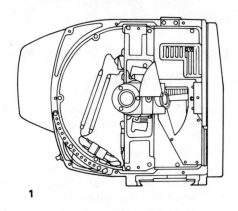

1

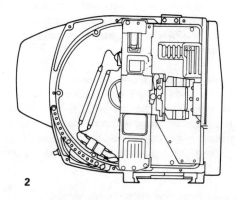

2

1. PANAFLEX camera with front removed showing spinning mirror reflex.
2. PANAFLEX camera with front removed showing pellicle mirror reflex.

TABLE SHOWING ⅓ STOP EXPOSURE INCREASES

Each step is a ⅓ stop exposure change									
22 >	20 >	18 >	16 >	14 >	13 >	11 >	10 >	9 >	
8 >	7 >	6.3 >	5.6 >	5 >	4.5 >	4 >	3 >	3.2 >	
2.8 >	2.5 >	2.2 >	2 >	1.8 >	1.6 >	1.4 >	1.2 >	1.1	

Changing the Ground Glass and Putting in a Cut Frame

Unless otherwise requested PANAFLEX cameras are normally supplied with an Academy or an Anamorphic ground glass depending upon the lens type ordered.

If it is an Academy frame it will show a reticle engraved to the ISO/ANSI CAMERA APERTURE dimensions. If it is known that the production is primarily intended for Theatrical release it will also be engraved with the 1.85:1 dimensions and if for Television with an inner rectangle showing the SAFE ACTION TELEVISION AREA. Ground glasses with many other markings, with combinations of markings, with and without center crosses and with special markings are available upon request.

The reticle markings of a PANAFLEX ground glass are mirrored to enable the PANAGLOW system to operate.

Removing and replacing the ground glass
To remove the ground glass from a PANAFLEX camera GENTLY withdraw the ground glass straight out of the camera, using a cotton swab or a lens tissue to protect the glass surfaces. Do not use pliers or any other metal tool to hold the ground glass as this causes chipping.

When replacing the ground glass insert it with the ground glass (dull) side towards the mirror.

Putting a cut frame into the viewfinder system
A frame cut from a previous take may be placed immediately in front of the ground glass for aligning SFX shots.

The selected frame must first be cut using a special frame cutter supplied by PANAVISION and then carefully inserted into a special ground-glass holder, marked with an "M," in front of the ground glass.

124

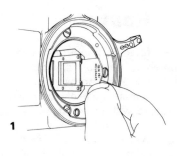

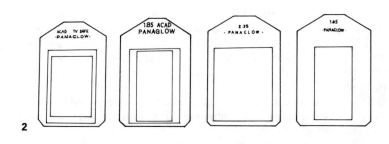

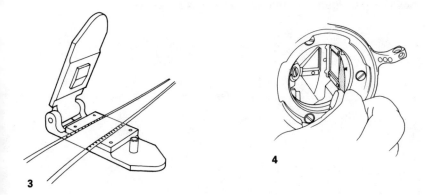

1. Removing a ground glass, 2. Various ground-glass markings, 3. Cutting a frame with a cutter, 4. Putting a cut into a ground-glass holder marked with an "M."

The Camera Heater

PANAFLEX cameras are fitted with internal heaters to keep the bearings warm in cold environments. The heaters, which are thermostatically controlled at 70°F (21°C) draw 2½ to 6 amps.

Whenever the internal heaters are likely to be used care should be taken in advance to ensure there will be an adequate supply of fully charged batteries or a battery eliminator available, bearing in mind the increased amperage required to power the heaters when they are in use. A yellow "heater-on" light to the right of the power supply socket on the rear of the camera indicates when the heaters are operating. A red LED warning light on the Annunciator Panel will indicate if the power supply is inadequate.

A 24 volt battery separate from that which powers the camera is plugged into the 2-pin LEMO socket to the right of the camera power socket. On the PLATINUM PANAFLEX a flashing red light, or on others a green/red light, between the two sockets indicates the condition of the heater battery when it is low, and an amber light indicates when the heaters are functioning.

On all models power for the magazine heater is connected internally to the camera heater supply. An amber light on the magazine indicates when the heaters are operating.

In conditions of severe cold it may be necessary to supplement the internal heaters with heated barneys (see pages 58-59).

The eyepiece optic may be kept clear from fogging by the PANACLEAR heated eyepiece. To operate, connect the power lead to the accessory power supply outlet on the magazine port cover.

126

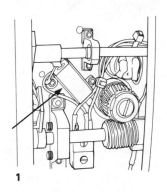

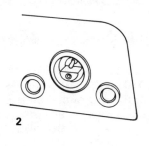

1

2

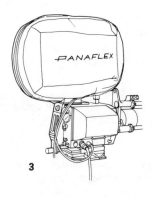

3

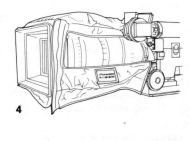

4

5

1. Heater units inside camera, 2. LEMO heater socket and green and amber warning lights, 3. Magazine heater barney, 4. Zoom lens heater barney, 5. PANACLEAR heated eyepiece.

127

PANAFLEX Magazines

Magazines for PANAFLEX cameras are available in 250, 500 and 1000', (75, 150 and 300m) capacities. The same magazine may be used on the top or the rear of the camera.

A special spacer unit is required when using a 250' magazine in the top position when a PANAVID video assist unit is fitted.

PLATINUM PANAFLEX magazines may be used on GII and other model PANAFLEX cameras, and vice versa, but may not be used with PANASTAR cameras which, because of their higher operating speeds, require special (500 and 1000') magazines. Note: The PLATINUM PANAFLEX magazine digital footage display will not operate when this type of magazine is fitted to a GII/GOLDEN PANAFLEX.

The 1000' and 500' magazines incorporate a mechanical "footage remaining" indicator which may be operated by pressing a lever on the rear of the magazine.

Unexposed film must be wound EMULSION-IN on a 2" center core.

Exposed film normally winds EMULSION-OUT, also on a 2" core, except in the case of PLATINUM PANAFLEX and PANASTAR *reversing* magazines (see below).

Locks on each spindle hold the cores firmly in position.

A single magazine lock, which is recessed to prevent accidental opening, locks the lid simultaneously in four places.

Reverse running magazines

Special reversing type magazines, which may be run in either direction, are available for the PLATINUM PANAFLEX and the PANASTAR cameras.

Reverse running magazines may be loaded in the normal manner, with the unexposed film loaded in the left-hand side, and the film run first forwards and then in reverse or the unexposed film may be loaded directly into the right-hand side of the magazine and the camera run in reverse from the beginning of the roll.

When a camera is run in reverse the film will take up EMULSION-IN.

Camera carrying handles

A magazine port cover, which incorporates a camera carrying handle and an outlet for the PANACLEAR eyepiece heater, must be fitted over whichever port is not in use.

A long top carrying handle, and a special rear magazine port cover to which it can be attached, is a supplied accessory for use with top-mounted magazines. An alternative top carrying handle and rear port cover is available for use when a PANAVID unit is fitted to the top of the camera.

128

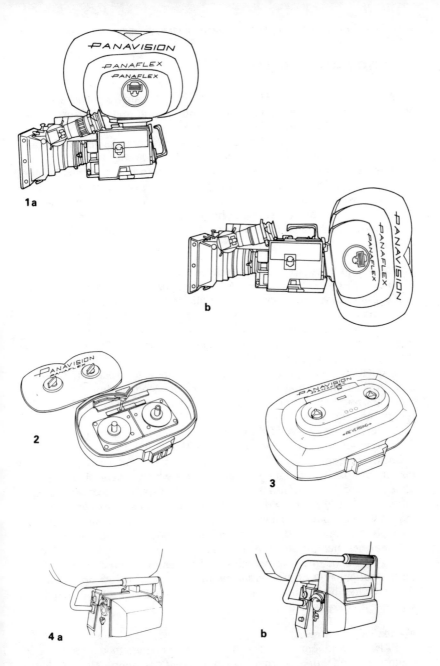

1.a & b. PLATINUM PANAFLEX showing 6 magazine positions/sizes, 2. Interior of normal magazine, 3. Exterior of reversing magazine, 4.a. Normal long camera carrying handle, b. Special long camera carrying handle for use when a PANAVID is fitted.

129

Magazine Systems

Electric contacts on PANAFLEX and PANASTAR camera magazine ports supply power to the magazine take-up motors and motor heaters and to the PANACLEAR and PANAFLASHER accessories.

Contacts which are common to all PANAFLEX cameras are: 2. Ground, 3. Magazine motor +, 5.24v + to PANACLEAR and other accessories.

Other contacts are used for supplying power to the magazine heaters and for the switching, reverse running and footage counter circuits.

When power is not connected to a magazine loose film on either the supply or on the take-up side can be tightened by pushing in the center of the motor and clutch covers on the back of the magazine and turning clockwise.

Footage and take-up running indicators

PLATINUM PANAFLEX and PANASTAR magazines incorporate an automatic shut-off switch which switches off the camera when there is 4' of film remaining on the PLATINUM, 5-10' on the PANASTAR. On the PLATINUM camera LED low-film warning lights on the annunciator panel will indicate when there is approximately 50' of unexposed film remaining in the magazine and when it is out.

PLATINUM PANAFLEX magazines incorporate an electronic "footage shot" indicator. This may be reset to zero by pressing the third reset button, the other two being used to reset the indicator slowly or quickly to predetermined footages.

On PLATINUM PANAFLEX magazines a blinking amber light indicates that the magazine is running. It is NOT a heater light.

On the magazines for GII and earlier PANAFLEX cameras the internal take-up motors incorporate a spiral indicator which shows when the take-up is running. If the camera is switched on, the spiral indicator runs comparatively slowly to show that film is running through the camera. If the indicator runs rapidly and continuously this signals that the film has passed through the camera and the camera has not been switched off.

Magazines incorporate a mechanical footage remaining indicator. It operates by means of a lever which may be pressed against the roll of film on the supply side of the magazine. In the case of reversing magazines this indicates how much film is in the left hand side of the magazine, whether it is exposed or not.

Magazine heaters

All PANAFLEX magazines incorporate electric heaters to keep the take-up motors in good working order in cold ambient conditions. On PLATINUM PANAFLEX magazines the heaters function automatically when power is applied to the camera heater. On earlier PANAFLEX magazines an amber light indicates when the heater is on.

130

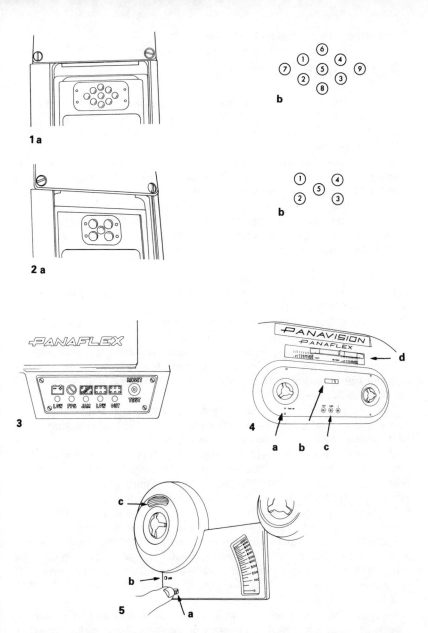

1. a. PLATINUM PANAFLEX magazine port electric contacts, b. Contact numbers,
2. a. Standard magazine port electric contacts, b. contact numbers, 3. PLATINUM
PANAFLEX annunciator panel, 4. PLATINUM PANAFLEX magazine indicators:
a. take-up running LED, b. digital footage shot indicator, c. digital indicator reset
buttons, d. manual footage remaining indicator, 5. Standard magazine
indicators: a. manual footage remaining indicator actuator, b. magazine heater
indicator LED, c. spiral take-up running indicator.

A very responsible job that must be done correctly.

Magazine Loading

Check that you have the correct type of magazine, i.e. PLATINUM PAN-AFLEX or PANASTAR or other, 1000, 500 or 250 ft capacity and, if it is a reversing type, if it is required to run a PLATINUM PANAFLEX or a PAN-ASTAR in reverse.

Check to see that no film remains in the magazine from a previous loading before removing a magazine door in the light. Hinge the door catches out of their recesses and turn both counter-clockwise to release.

Transfer the plastic center core from the supply side of the magazine and place it on the take-up spindle with the film slot facing counter-clockwise. If there is not a spare core in the magazine one must be provided.

Push back the top lock of the supply side spindle.

Things to be done in the dark

Extinguish the room light or seal up the changing bag and remove the unexposed film from its can.

For normal use set the film on the left side of the magazine with the film coming off the roll in a clockwise direction.

In the case of reversing magazines the unexposed film must go into the left-hand side of the magazine if the first run is in a forward direction or into the right-hand side of the magazine (and unwind anti-clockwise) if the first run is in reverse.

Take the end of the film between the thumb and first two fingers to curve slightly and push between the two front rollers until it comes out of the magazine throat (rear two for reversing magazines).

Move the roll of film over, slide the center core onto the spindle, and lock down. Pull the end of the film out of the magazine and press down on the spindle until an audible click is heard, indicating that the spring-loaded key is securely fitted in the keyway of the core.

Pull down approximately 2-3' of film, bend the end of the film again and pass between the rear magazine rollers until it is inside the magazine. Pull the film to the right hand side of the core, locate the end in the core slot, turn the take-up core counter-clockwise and wind on two or three turns of film. (Left hand side and clockwise on reversing magazines.) For normal forward-running magazines the film path is in the form of a "99" and for reverse-running magazines it is "9P."

Re-install the magazine lid (with locks vertical and turned fully counter-clockwise) by placing it flat on the body of the magazine (it is not necessary to engage any lugs) until it seats properly. Turn both locks clockwise and snap down into their recesses to double lock in position.

132

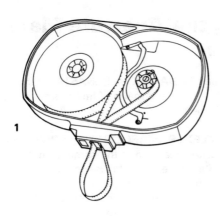

1

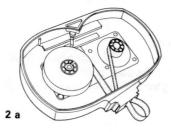

2 a

b

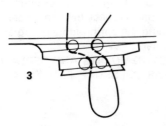

3

4

2. *Normal* magazine showing the normal "99" film loading path, 2.a. *Reversing* magazine showing the "9P" loading path to run the film in a *forward* direction, 2.b. Reversing magazine showing the "9P" loading path to run the film in a *reverse* direction, 3. The route through the light trap rollers, 4. Magazine lid locks.

133

Checking the Aperture Plate and Matte

Before operating any PANAFLEX or PANASTAR camera check that the aperture plate is clean, that the matte (if fitted) is appropriate to the shoot and that the gate is clean.

Removing PANAFLEX, PANAFLEX 16 and PSR aperture plates
To remove the aperture plate, inch the camera until the pull-down claw is at the bottom of its stroke, after it is disengaged from the film, and the registration pins are fully engaged.

Pull out the spring-loaded register pin retraction knob and slide it back to clear the register pins from the film perforations.

Turn the gate top dog lock clockwise and turn the bottom thumb lock counter-clockwise. Hold the bottom lock horizontally between the thumb and index finger, and wiggle to remove the gate plate from the camera movement.

Removing a PANASTAR aperture plate
On a PANASTAR camera, inch the camera until the pull-down claws are fully engaged and the register pins are in the back position.

Retract the pull-down claws by pulling on the spring-loaded retraction knob and pushing it downwards.

Checking the matte
To check the matte for size and cleanliness the gate must be removed.

Mattes for Super 35, Academy, 1:85 and 1.66:1 are available on request. Full Aperture and Anamorphic formats do not require a matte.

Replace the aperture plate by holding the bottom lock horizontally and pushing the plate onto the top locating pin. Push the bottom lock down and set the top dog lock by pulling it out and turning counter-clockwise.

Occasionally take out the rear pressure pad to check for cleanliness and freedom from abrasion. Remove by turning the spring-loaded dog at the rear outward. Reverse the process to replace.

GREAT CARE MUST BE TAKEN WHEN REMOVING AND REPLACING THE APERTURE PLATE AND THE REAR PRESSURE PAD TO AVOID DAMAGE.

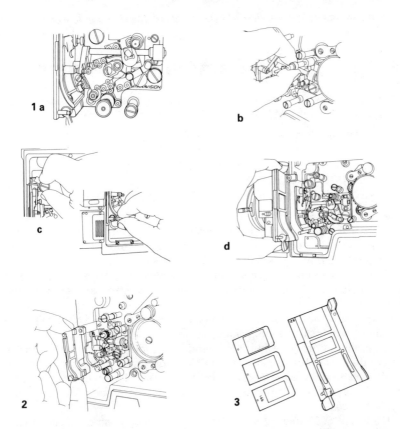

1. Removing a PANAFLEX aperture plate: a. Set the pull-down claws to bottom of stroke, b. Pull out the register pin lock and slide it back, c. Release the gate locks, d. Remove the aperture plate CAREFULLY, 2. Removing a PANAFLEX 16 aperture plate, 3. PANAFLEX 35mm aperture plate and various mattes.

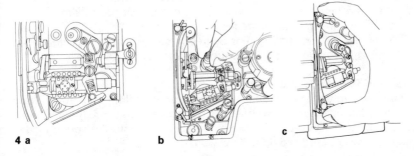

4. Removing a PANASTAR aperture plate: a. Set pull-down claws into the film with the register pins in the out position, b. Pull out the spring loaded knob and push it downwards to retract the pull-down claws, c. Release the aperture plate and remove CAREFULLY.

135

With PANAVISION you can choose a short and tall or a long and low camera configuration.

Preparing the Camera for Loading

On the PLATINUM PANAFLEX set the Digital Indicator to the desired display by pressing the MODE switch. The mode will change as follows:

CAMERA SPEED (fps) + FOOTAGE (exposed)
CAMERA SPEED (fps) + METERAGE (exposed)
CAMERA SPEED (fps) + SHUTTER ANGLE (°)
TIME CODE TIME OF DAY
TIME CODE USER BITS

(See page 93 for illustrations)

"FIL" in the bottom left corner indicates a gelatin filter holder is in place.

If an extension eyepiece is fitted, semi-tighten the friction lock at the viewfinder "elbow" where "LOCK" is engraved.

Release the eyepiece leveller link arm by squeezing together the pair of levers below the underside of the extension eyepiece and stow on the rest.

Disconnect the PANACLEAR eyepiece heater cable if fitted, raise the eyepiece and lock it in an upright position.

Depending upon whether it is intended to mount the magazine on the top or the rear of the camera, remove the appropriate magazine port cover and check that the other port is covered securely. An alternative rear port cover is available which incorporates an attachment point for a long carrying handle.

If a 250 or 500' magazine is to be used in the top position together with a PANAVID video-assist system a magazine spacer unit must be fitted.

Close the top and bottom sprocket keeper rollers.

Setting PANAFLEX, PANAFLEX 16 and PSR cameras

Open the camera door and inch the movement until the pull-down claw is at the *bottom* of its stroke, just after it comes out of the film, and the *registration pins are fully engaged.*

Withdraw the registration pins by gently pulling out the spring-loaded retraction knob and sliding it towards the rear of the camera.

Setting PANASTAR cameras

Open the camera door and inch the movement until the pull-down claws are fully *into* their pull-down stroke and the *register pins are fully withdrawn.*

Withdraw the pull-down claws by gently pulling out the spring-loaded retraction knob and sliding it *downwards.*

An optional safety measure

The power supply may be disconnected before threading film as a safety measure.

136

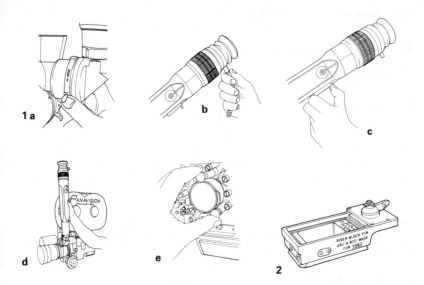

Preparing all cameras for loading: 1. a. Eyepiece friction lock to be set no more than semi-tight to enable the eyepiece to be moved clear of the camera door, b. PANACLEAR disconnected, c. Eyepiece leveller released, d. Eyepiece raised, e. Top and bottom keeper rollers closed, 2. 250/500' magazine spacer unit.

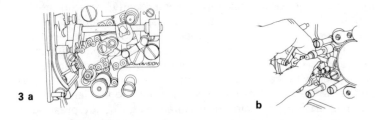

3. Setting PANAFLEX and PSR cameras: a. Set movement with pull-down claws at bottom of stroke, b. Registration pins slid back clear of the film path.

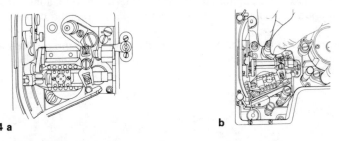

4. Setting a PANASTAR camera: a. Set movement with pull-down claw into pull-down stroke, b. Pull-down claws slid back clear of the film path.

137

Threading the Camera— Lacing the Film

Inch the camera until the pull-down claw is at the bottom of its stroke.

Check that the magazine is loaded with the appropriate film stock and pull out a very short loop from the supply (front) side of the magazine.

Rest one end of the magazine on the camera, feed the film loop through the magazine port and lock the magazine securely in position.

Special note: WHILE THREADING THE CAMERA *DO NOT PRESS THE INCHING KNOB* ON THE REAR OF THE CAMERA UNTIL THE FILM ON THE TAKE-UP SIDE HAS BEEN SECURELY LOCATED ON THE SPROCKET AND THE BOTTOM SPROCKET KEEPER HAS BEEN CLOSED. (Pressing in the inching knob activates the magazine motor and this is not a good thing to do until the film has been secured on the main sprocket.)

Lacing the film

Pull about 8″ (20cm) of film from the front compartment of the magazine and stretch it towards the bottom left hand corner of the camera.

Open the top and bottom sprocket keepers.

Thread the film through the camera exactly as shown on the threading diagram on the inside of the camera door. Double check that it is correct.

Check that the film on the take-up side is properly seated on the underside of the sprocket and close the bottom sprocket keeper.

Set the bottom loop so that it just clears the bottom of the camera.

Press a sprocket hole onto the perforation locating pin situated just above the aperture plate. This will ensure that the perforations will be correctly aligned with the registration pins. At the same time gently press on the edge of the film to ensure the film is fully into the camera.

Set the registration pins into the perforations by gently pressing the boss at the base of the retraction pin. (Note: Pressing the base of the retraction pin has a more positive feel than pushing from the top). If it does not go easily recheck the perforation alignment.

Set the top loop by pulling the film off the perforation locating pin. The top loop should be set to clear the locating pin as per the drawings on page 141.

Engage the film on the top of the sprocket and close the top keeper.

If it has been disconnected, re-connect the camera power supply.

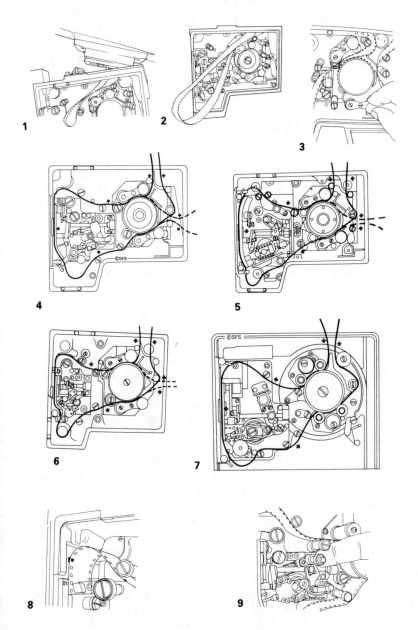

1. Pull out a short loop of film and rest the magazine on the camera, 2. Pull loop of film to bottom L.H. corner of camera, 3. Open top and bottom sprocket keepers, 4. PANAFLEX threading diagram, 5. PANASTAR threading diagram, 6. PANAFLEX 16 threading diagram, 7. SUPER PSR threading diagram, 8. Press film on perforation locating pin, 9. Press in register pins by base of retraction pin.

139

Threading the Camera—Checking That It Runs Properly and Quietly

On the PLATINUM PANAFLEX press the inching switch on the inside of the camera to take up any unwound film in the magazine. On earlier cameras slide away the small lever behind the rear inching knob and press in the knob to take up any excess.

Inch the camera through a complete cycle to check the lacing. Check that the top loop does not touch the perforation locating pin and that the bottom loop does not become too tight or touch the bottom of the camera. If a loop is not correct open the corresponding sprocket roller and adjust as appropriate.

Set the switch at the rear of the camera to ON. If the camera door is open the camera will not run unless the micro switch at the bottom right-hand corner of the film compartment is depressed. (Not PLATINUM.)

Setting the camera for maximum quietness

All PANAVISION cameras (except the PANASTAR) incorporate a pitch control to align the perforations with the registration pins at the end of the pull-down stroke to ensure quiet running.

To run the camera at its operating speed while at the same time adjusting the pitch of the pull-down claws right-handed people will find it easier to hold the pitch adjustment control knob with the right hand and use the left hand to depress the micro-switch at the bottom right-hand corner of the camera body. For left-handed people it is the reverse. On the PLATI-NUM simply press the RUN button.

With the camera running turn the pitch control knob clockwise and counter-clockwise until the perforation noise is minimized. This adjustment may have to be made after every re-load to ensure that the camera runs as quietly as is possible for a camera which has full fitting registration pins and maximum image steadiness to do. It may even be advisable to check the pitch setting during the course of a single roll of filmstock.

In addition to adjusting the pitch of the camera to locate the perforations in relation to the register pins, PANAVISION and its representatives world-wide have a special tool available which enables them to adjust the stroke to optimize the amount of film which is pulled down each time. Camera-men using filmstock which may be slightly different to normal are advised to bring in a sample roll so that the stroke may be set to suit.

Let go the micro switch (or the RUN button), switch OFF the camera and close the camera door.

Note: It is very important, especially on the PLATINUM PANAFLEX, to double check, before closing and locking the camera door, that the camera has switched OFF at the main camera switch. The camera is so quiet that it is possible to run a whole roll of film through without realizing it.

140

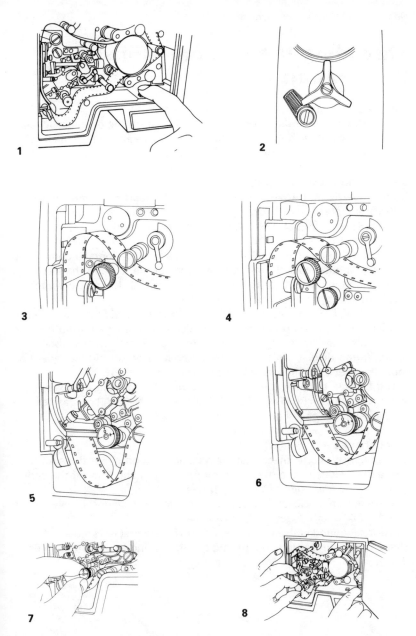

1. PLATINUM PANAFLEX internal inching switch, 2. Rear inching knob with inhibiting lever to prevent accidental operation, 3. The correct top loop with claw at top of stroke, 4. The correct top loop with claw at bottom of stroke, 5. The correct bottom loop with claw at top of stroke, 6. The correct bottom loop with claw at bottom of stroke, 7. Adjusting the pitch control, 8. Adjusting the stroke.

Threading the Camera—Final Preparation

Pull the eyepiece down to the shooting position.

If the camera is mounted on a PANAHEAD attach the eyepiece leveller link-arm to the underside of the extension eyepiece and unlock the friction lock at the eyepiece elbow.

Setting the footage counters

On the PLATINUM PANAFLEX set the digital footage/meterage counter to zero by simultaneously pressing the buttons marked with up and down arrows. Alternatively, these buttons may be used individually to preset the counter to any particular starting point.

On the GII and earlier PANAFLEX cameras zero the footage counter by pressing simultaneously the micro switches at the top and the left hand side of the digital footage read-out. Press only the top micro switch to recall the footage shot since the last resetting.

On PANASTAR *reversing* magazines there is a Digital Frame Counter to enable the film to be rewound to a particular frame (± 1 frame) for SFX work. This may be reset by pressing the button set directly above the digital frame display. Note: This frame counter is not bidirectional and must be reset each time when reverse running.

The footage remaining in a magazine may be checked manually by a sliding actuator on PLATINUM PANAFLEX and PANASTAR magazines and by a small lever on GII and earlier magazines.

Final checks

It is good practice to check there is no slack film on either the supply or the take-up sides of the magazine by pressing in and turning the centre knobs on the back side of the magazines.

In humid and cold ambient conditions connect power supply to the PANACLEAR heated viewfinder eyepiece.

Before shooting with a particular camera for the first time always double check the shutter setting, the fps rate and the behind-the-lens gelatin filter, if any.

142

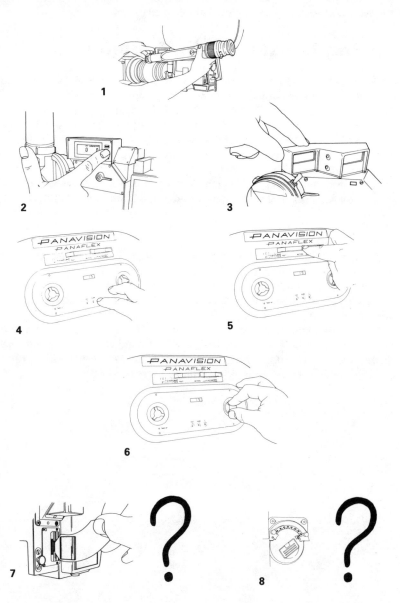

1. Reposition eyepiece and reconnect the eyepiece leveller and PANACLEAR,
2. Reset PLATINUM PANAFLEX footage counter by simultaneously pressing the
up and down buttons, 3. Reset the GII footage counter by simultaneously
pressing the top and left-hand side buttons, 4. Resetting the footage counter on
a PLATINUM PANAFLEX magazine, 5. Checking the footage remaining on a
PANAFLEX or PANASTAR magazine, 6. Tightening the roll of film inside a
magazine, 7. A reminder that the gelatin filter holder must be checked, 8. A
reminder that the adjustable shutter quadrant must be checked.

143

Fitting Lenses to the Camera

All PANAFLEX and PANASTAR cameras, together with PSR cameras and PANAVISION versions of Arriflex and Mitchell cameras are fitted with the same strongly designed PANAVISION lock ring lens mount.

To mount a lens to the camera remove the lens port cover from the camera (if fitted), and turn the lens locking ring fully counter-clockwise.

Remove the rear cover from the lens and offer up the lens to the camera with the locating pin in the downwards position, slide the lens firmly and squarely into the camera and turn the locking ring fully clockwise to secure.

Long focal length and zoom lenses may require additional support from the iris rods.

Range extenders

Range extenders are available for certain telephoto and zoom lenses which increase their focal length by a factor of 1.5 and 2 times.

They reduce the effective aperture by the square of their magnification. Thus a 2× range extender working with a 400mm T4 telephoto lens converts it to an 800mm working at T8.

Range extenders certainly do not improve the optical quality of a lens and may significantly impair it. They are best used with the lens stopped down at least two stops. The image degradation sometimes looks worse through the viewfinder than it does on the screen.

144

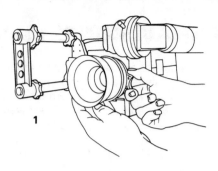

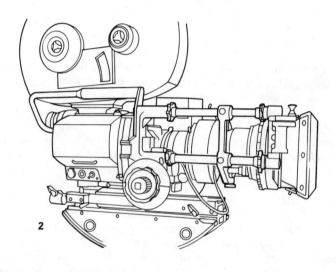

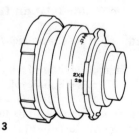

1. Fitting a lens to a camera, 2. Long lens with support, 3. Range extender unit which may be used with zoom or long focus lenses.

145

Testing PANAVISION Lenses

Every lens is unique and before any production commences PANAUSERS are encouraged to thoroughly test every lens they are taking with them.

Bear in mind that almost every lens performs best one or two stops down from maximum aperture and performs least well at minimum aperture. Some lenses will focus on slightly different focus marks at full aperture compared to two stops stopped down. Some zoom lenses have a focus shift over the zoom range, have better performance at some focal lengths than others and change their image size as the focus setting is changed. Some lenses appear to have more or less depth of field than others or more or less depth in front or behind than may be expected.

Lenses incorporate many different types of glass and thus some may be warmer than others. The color of the lens coating is nothing to go by; it is the color of the image on the film that matters.

All lens design is a compromise and where one lens has less of any of these deficiencies than another then it may be at the cost of performance elsewhere. For this reason Camera Assistants are given every opportunity to get to know the capabilities of their PANAVISION lenses before a shoot in order to be able to exploit each one to its maximum.

Lenses should be tested for a *combination* of definition *and* contrast, where it focuses (both by tape and by eye) at different apertures, different distances and, with zoom lenses, at different focal lengths, for veiling glare over bright highlights, distortion of vertical and horizontal lines, depth of field, color, etc.

For optimum performance assessment Camera Assistants are invited to put their lenses through their paces on PANAVISION's Optical Test Laboratory where they may see exactly what each lens does on an MTF (which measures a combination of contrast and definition) and colorimetry test benches. (Please telephone PANAVISION's optical department to make an appointment beforehand.)

Notes on understanding MTF (Modulation Transfer Function) graphs

All graphs illustrated on the right are for on-axis (i.e. center of the optical axis) only.

The Diffraction Limit is the natural limit of contrast at any resolution at a given aperture of *any* lens. The Average Resolution for Cinematography is as defined by the International Standards Organization (ISO).

It is worth noting that diffraction *alone* limits the maximum achievable contrast at any resolution (no matter what lens) and that diffraction becomes a greater limitation as the apertures become smaller, i.e. T8, T11, T16, etc., and the focal lengths become shorter, i.e., 20mm compared to 250mm.

MTF is similar to contrast when resolution is specified.

146

 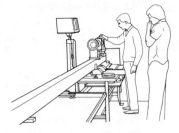

1. PANAVISION camera test room, 2. PANAVISION's MTF test bench.

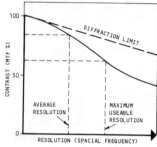 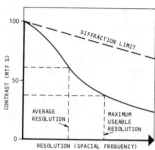

3. MTF graphs of a typical prime lens at T2.8 (left) and at T1.3 (right). (Note that there is a significant increase in both contrast and resolution as the lens is stopped down.)

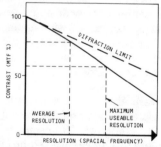 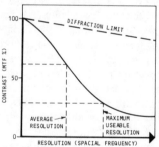

4. MTF graphs of a PANAVISION PRIMO and of a typical prime lens. (Both at T2.) (Note the considerably superior contrast/resolution characteristics of the PRIMO lens.)

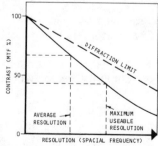 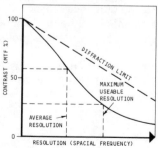

5. MTF graphs of a zoom lens (at T4) at the wide-angle and at the telephoto settings. (Note the improved performance at the wide-angle end of the zoom range.)

147

Focusing PANAVISION Lenses

Every PANAVISION lens is calibrated individually to achieve optimum optical performance. *Lenses* are calibrated against a standard flange focal depth of 2.2500in, 57.15mm, at an aperture of T2.8 (or at the minimum stop above T2.8) and an index mark is engraved on the lens to indicate the best possible image at that aperture. Lenses are calibrated at all marked focus distances.

Lenses are then further calibrated at maximum aperture and any difference from the T2.8 setting is noted. If there is a significant difference between the position of the T2.8 and the full aperture mark a second (shorter) mark is engraved and this is colored blue. The apertures to which it refers are also colored blue.

This extensive calibration enables PANAVISION to spread out its lens scales to a much greater extent so that Focus Assistants can achieve maximum precision when setting focus. Unfortunately this also means that any aperture-related focus shift tends to be more noticeable than on a compressed lens scale which hides focus shift. PANAVISION believes that the need to have a separate index mark for the very wide apertures only is a small price to pay for large lens barrels with big focus scales.

In general it is always preferable to use a tape measure, rather than the ground glass, to determine the focus settings for wide-angle lenses at apertures greater than T2.8. This rule applies to all wide-angle, wide-aperture lenses on all cameras and is because the aperture of the viewfinder system of PANAFLEX, and all other cameras, is not as wide as a wide aperture motion picture lens and thus eye focusing will not necessarily be correct at wide apertures.

Special care should be taken by Camera Assistants when eye focusing lenses which have alternative wide aperture focus markings (engraved blue and considerably shorter) for use at full aperture. When tape focusing such a lens at a blue-colored aperture, the focus distance settings should be set to the blue-colored index mark. When focusing these lenses by eye at apertures which are colored blue the correct distance will be that which is against the blue index mark.

Checking infinity by auto-collimator can also be inaccurate and misleading, especially with wide-angle, wide-aperture and long focal length lenses.

Note: Before setting lens focus by eye the focus of the ground glass through the viewfinder must first be set to suit the person who is doing the focusing.

148

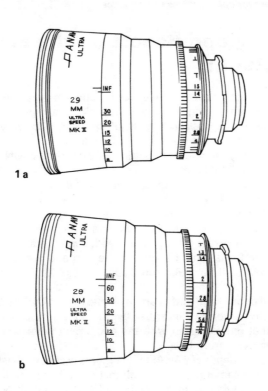

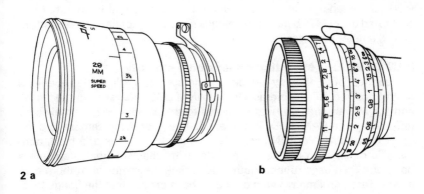

1. Lens scale showing the normal (yellow) and full (blue) aperture index lines (a.) set to the optimum infinity position for T.2 and smaller apertures, and (b.) set to the optimum infinity position for T1.3 and T1.4, 2.a. PANAVISION spread-out lens scale, 2.b. A typical compressed lens scale. Note how focus shift may be hidden by compressing the lens scale.

149

Checking and Resetting the Flange Focal Depth

To ensure the best possible image quality and the reliability of the engraved scale when setting focus with the aid of a tape measure, it is imperative that the flange focal depth of the camera be very accurately set with the aid of a depth-measuring micrometer.

The correct flange focal depth setting of all PANAVISION *cameras* is 2.2488 ± .0001" (57.1195mm).

The only accurate method of checking the flange focal depth of a PANAFLEX camera is by the use of a depth gage and a special steel plate which is held against the camera aperture plate in place of the film. By comparison an auto-collimator is a comparatively inaccurate instrument which may give incorrect settings because it cannot make allowance for the fact that the focus must be set *into* the film and not on the surface, and that the film will normally run 0.0001" to 0.0003" forward from the pressure pad.

Flange focal depth gage outfits are available from PANAVISION Inc. and its representatives worldwide for use by trained technicians and it is recommended that one be taken with the equipment whenever the hire is likely to be far away from base.

The flange focal depth of PANAFLEX cameras is set by adjusting special screws set on the camera. It is not a job that should be undertaken by any Camera Assistant without special training by a PANAVISION technician. Training is available on request.

The ground glass should be set at 2.250" by collimation using a short focal length lens (no longer than 30mm) which has had its infinity mark calibrated on an MTF bench or by means of a collimator having an aperture of 1" diameter or larger.

PANAFLEX lens mounting dimensional stability

A great deal of mythology has been generated by stories of "rubber lens mountings" on quiet cameras. On all PANAFLEX cameras both the lenses and the movements are hard mounted to the camera body. It makes no difference on a PANAFLEX if the lens is heavy or if the camera is tilted up or down. PANAVISION uses special mountings made of a hard, non-metallic material to isolate the pull-down mechanism from the camera body and then uses rubber mounts to isolate the motor from the mechanism. Tests, and many years of usage, have proven that the flange focal depth of PANAFLEX cameras is not affected out of tolerance by steep tilts.

150

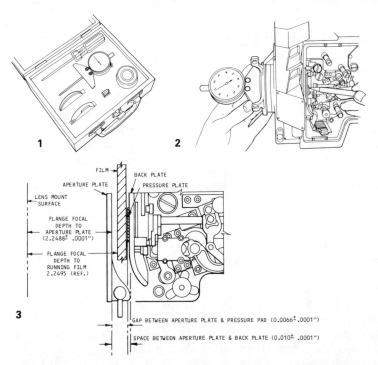

FILM

BACK PLATE

APERTURE PLATE

PRESSURE PLATE

LENS MOUNT
SURFACE

FLANGE FOCAL
DEPTH TO
APERTURE PLATE
(2.2488± .0001")

FLANGE FOCAL
DEPTH TO
RUNNING FILM
2.2495 (REF.)

GAP BETWEEN APERTURE PLATE & PRESSURE PAD (0.0066± .0001")

SPACE BETWEEN APERTURE PLATE & BACK PLATE (0.010± .0001")

1. Flange focal depth gage set, 2. Method of using a flange focal depth gage on a camera. Note the use of a pencil to hold the gage plate in position behind the aperture plate, 3. Critical lens/film/camera distances.

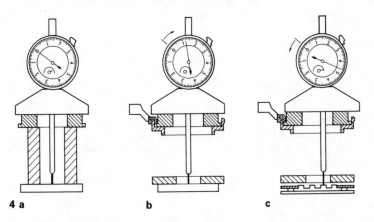

4 a   b   c

4. Flange focal depth checking procedure: a. Set the dial gage to zero with the aid of a 2.2500" gage block, plate and adaptor flange, b. With a gage plate behind the aperture, and using the adaptor flange, check the flange focal depth of the camera. It should read −.0012" ± .0001" on the dial, c. Check the depth of two center rails of the pressure plate. They should register +0054" on the dial.

151

Where the Director of Photography does much of his image control.

Matte Box Features

A standard matte box is supplied with every PANAFLEX camera. It attaches to the camera by two iris rods (called "support bars" in the UK) set one above the other on the motor side of the lens.

It takes two 4 × 5.650" rectangular filters and has a 5.5960" diameter circular retainer ring which can take a 138mm Polarizing Filter or other diameter circular filter or a 4½" diameter reducing ring which can also be fitted with a "snout" which can be used with very wide angle prime lenses to clear the viewfinder optical system.

The top is especially strengthened to be fitted with a sunshade flap.

A smaller, lightweight, "spherical" matte box, which is particularly suitable for use with physically small lenses or when hand-holding the camera, is available on request. It incorporates holders for two 4 × 5.650" rectangular filters and one 4½" diameter circular filter holder.

Both matte boxes are hinged to allow them to be swung clear for lens changing. To release, lift spring-loaded knob above the hinge and swing the matte box forward.

For the 20-100mm Super Zoom a special Wide-Angle Matte Box which takes only one 4 × 5.650" filter is supplied.

All matte boxes are supplied with slip-on, lightweight, plastic mattes which fit over the front of the sunshade and have cut-outs to suit various focal length lenses. These mattes minimize the amount of stray light entering the matte box from the front. Foam rubber "Donut" rings or bellows are supplied with all lenses. They fit between the lenses and the rear of the matte box to prevent stray light from entering from behind which could reflect off the rear of the filters.

Tilting matte box
A matte box which enables the filter to be tilted forwards to eliminate distracting multiple reflections caused by automobile headlights, etc. is a very useful accessory when filming car chases at night.

Multi-stage matte box
A multi-stage matte box which takes any number of sliding and rotating 6.6" sq. filters, plus tilting and mechanically operated sliding filters, is an available extra.

152

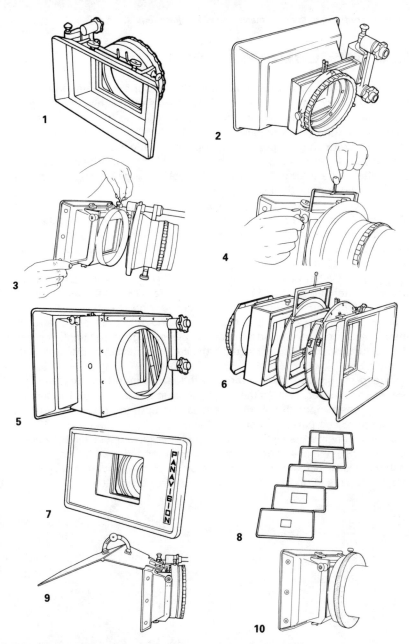

1. Standard matte box, 2. Spherical matte box, 3. Swinging a matte box open to facilitate lens changing, 4. Adjusting a sliding graduated filter tray, 5. Tilting matte box, 6. Multi stage modular 6.6″ sq. matte box, 7. Matte box matte in position, 8. Set of matte box mattes, 9. Extension sunshade flap, 10. Donut in position between a lens and the rear of a matte box.

153

Iris rods are there to support matte boxes and long lenses.

Iris Brackets and Rods

Two types of iris rod brackets are available to support a matte box and long and heavy lenses when appropriate. A short, lightweight type is available for use when hand-holding and a long type is available for use when supporting heavy lenses.

To fit iris rods that support a matte box only, slide the short bracket into the vertical slides situated on the motor side of the lens. The iris rods and the matte box are locked in position by locking rings which should only be tightened "finger tight".

PANAFLEX cameras are normally supplied with three pairs of iris rods 5", 7" and 9" long. The pair of the most suitable length for a particular lens should be selected and fitted onto the iris rod bracket. PLATINUM and GII PANAFLEX cameras may be supplied with iris rods made of carbon fiber to give added lightness.

To support a heavy lens fit the iris rod bracket to the camera (but do not completely tighten) and select a pair of iris rods of suitable length.

Fit the lens and tighten the lens mount locking clamp in the normal manner. Initially fit only the top iris rod (remove bottom one if already fitted) and slide lens support bracket into place. Engage and lightly tighten the captured locking screw to attach the support bracket to the underside of the lens.

Thread the second iris rod through the lens support bracket and into the iris rod bracket. Tighten the iris rod bracket attachment to the camera, and the lens support bracket to the lens and both iris rod locking rings, checking that no part is under any strain.

Note: In an emergency the short iris rod bracket can be used to support long and heavy lenses but it is not a recommended practice. In such an instance it is particularly important not to tighten the bracket to the camera before the lens support and both iris rods have been fitted.

154

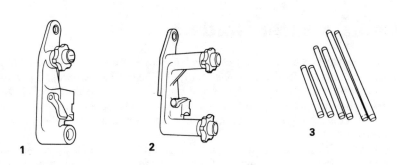

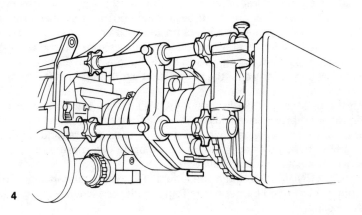

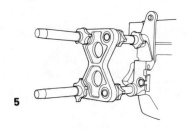

1. Short iris rod bracket, 2. Long iris rod bracket, 3. Set of iris rods, 4. Iris rods supporting a long and heavy lens, 5. Iris rod offset bracket for use with a 6.6" matte box.

Gelatin Filter Holders

In addition to places for three or four filters in the various PANAVISION matte boxes there is also a provision on all PANAFLEX cameras to place a gelatin filter just in front of the film plane and on certain PANAFLEX zoom lenses to place a gelatin filter behind the rear optical element.

PANAFLEX cameras are supplied with a box containing 12 gelatin filter holders.

Before a shoot the Camera Assistant should very carefully mount a selection of gelatin filters into the holders as requested by the Director of Photography. PANAVISION Inc. and its distributors worldwide can supply a gelatin filter punch for cutting filters to shape. Be careful to keep dust, finger marks and other blemishes off the surface of the filters. The use of a PANAVISION filter punch makes this task easier.

The use of a gelatin filter between the lens and the film affects the lens back focal distance by about 1½ thousandths (.0015) of an inch. This is not likely to have a deleterious effect on the focus of any lens unless it has a particularly short focal length and/or wide aperture. If the Director of Photography plans on using gelatin filters he may wish to request that PANAVISION technicians alter the flange focal depth setting of the camera accordingly.

To fit the gelatin filter holder into a PANAFLEX camera, slide back the dust/light/sound-proof cover below the viewfinder tube, insert the filter holder inwards and upwards, close the cover slide and remind the D.P. that a behind-the-lens filter is in place so that he can make appropriate allowances in his exposure calculations.

On the PLATINUM PANAFLEX the letters "FIL" will automatically show in the bottom left hand corner of the Digital Indicator whenever a gelatin filter is in place.

Cutter punches are also available for the circular filters that can be placed behind most PANAZOOM zoom lenses.

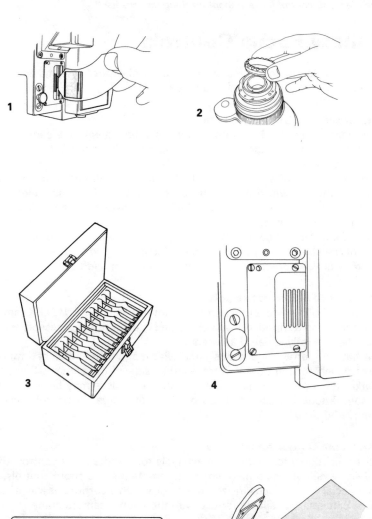

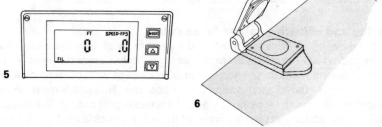

1. Putting a "behind the lens" gelatin filter into position, 2. Fitting a filter onto the rear of a zoom lens, 3. Set of gelatin filters in a box, 4. Closed gelatin filter door, 5. "FIL" indicator on PLATINUM PANAFLEX display, 6. Gelatin filter cutter.

Follow Focus Controls

A variety of follow focus, iris control and zoom control systems are available for PANAFLEX cameras and lenses.

The Standard Follow Focus unit
The Standard Follow Focus unit, which is supplied with the camera, has a single knob which can normally be operated from the left side of the camera only.

In addition, a follow focus handle, for fine control, and a follow focus extension cable, which may be used on its own or in combination with the handle, are provided to enable the focus assistant to stand a little way back from the camera.

The Standard Follow Focus unit may be operated from the right-hand side of the camera by using the follow focus extension.

It may also be operated by the camera operator's left thumb.

The Super Follow Focus unit
The Super Follow Focus unit is intended for use when a PANAFLEX camera is mounted on a tripod, dolly or crane and is an optional accessory which must be specially ordered as needed.

It has two large hand knobs fitted with magnetically attached marker discs which can be operated from either side of the camera. Since the marker discs are interchangeable a number of discs may be marked up to suit various lenses. The knob on the motor side of the unit may be removed if required.

Electronic follow focus units
Electronic follow focus units are available for remote focus control. They are also available in combination with remote iris and zoom controls.

A special remote focus/iris/zoom control, which incorporates a lens scale video camera, is available for use with the Louma camera crane.

Fitting and adjusting follow focus units
Both follow focus units are fitted to the camera by sliding them onto the short rod which protrudes from the camera body below the lens. After fitting, the unit must be pushed upwards to engage with the lens gearing until a spring-loaded catch snaps it into position. This catch must be released every time the lens is changed. To release, pull out the flat-topped knob at the bottom left hand corner of the left hand side of the camera.

The meshing of the follow focus gears may be adjusted by means of a small set screw situated near the follow focus unit attachment point.

Removable (magnetic) lens calibration and focus marking disks may be marked up to suit any lens and fitted to either type of follow focus unit.

158

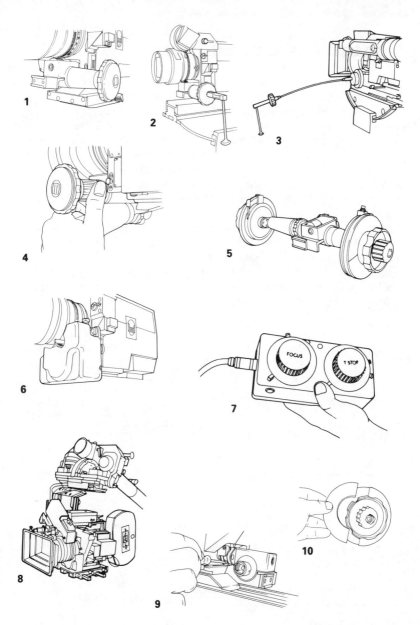

1. Standard follow focus control, 2. Follow focus with handle fitted, 3. Follow focus with long extension and handle fitted, 4. Using a thumb to adjust focus while hand-holding, 5. Super follow focus unit, 6. Electronic type follow focus, lens unit, 7. Electronic type follow focus, hand unit, 8. Special Louma camera crane lens control unit, 9. Adjusting follow focus gear meshing, 10. Changing follow focus lens calibration disks.

Remote Iris Control Systems

Various Remote Iris Control units are available to control zoom and fixed focal length lens apertures by use of the remote control cable and handle.

The Remote Focus Iris and Control unit is an electronic system designed to control focus and exposure of normal focal length Spherical lenses from a distance. It consists of a dual motor unit, which fits below the lens and normally draws its power from the zoom lens power outlet on the motor side of the lens, and a remote unit with separate controls for each function. The two units are connected together by an extension cable.

The system may also be operated from an independent battery by plugging a 24v DC supply directly into the remote unit.

Before fitting the lens unit both fuctions should be set to their maximum reach in a counter-clockwise direction, the lens focus set to infinity and the lens apertures set to full aperture.

Engage the lens unit, leaving a little back lash, and check to see that it runs freely. Rotate the knobs and mark both sets of calibrations onto the white magnetic discs.

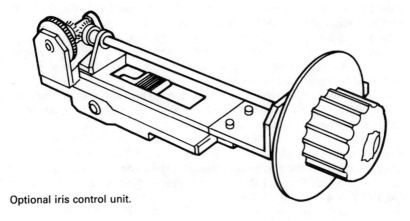

Optional iris control unit.

Zoom Controls

All PANAVISION zoom lenses may be operated either manually or electronically using an Electronic Zoom Control unit. A clutch on all lenses must be slid across to select whichever means of control is required (except for those lenses which have a "collar" motor).

When operating in the electronic control mode the zoom control hand unit (zoom gun) must be connected to the zoom motor by the yellow cable supplied. The zoom motor must be connected to a 24v DC power supply by one of the black cables supplied.

When drawing power from the zoom control power outlet on the front of the PANAFLEX camera use the short cable and when operating off an independent battery use the longer cable.

A switch marked IN and OUT on the back of the hand-held unit determines the direction of the zoom—IN to long focal length, OUT to wide angle. A knob on the top selects the operating speed and a finger grip switches the unit ON and OFF. A button on the top of the unit, marked FAST RETURN, may be used to reset the zoom with the minimum of delay. Soft start and stop movements may be made by using the finger grip control and speed selection knob simultaneously.

Zoom control parking bracket

A special bracket which fits into the cave on the side of the camera, is available to hold the zoom control unit when not in use or when it is more convenient to use attached to the camera setup.

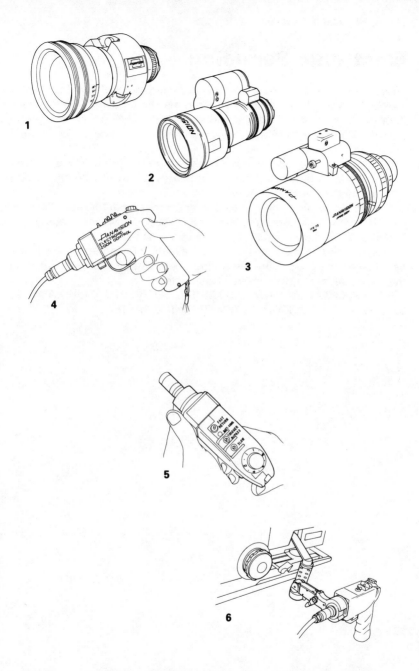

1. Collar type zoom motor, 2. Piggy-back type zoom motor, 3. PANAVISION PRIMO zoom lens and motor, 4. Zoom control hand unit (Zoom gun), 5. Zoom controls in detail, 6. Zoom control parked on holder unit.

163

Electronic Servicing

Almost all of the principal electronic components of the PLATINUM PANAFLEX, including the on-board computer chips, are mounted on three (four in the case of the GII and earlier PANAFLEXES) easily changeable plug-in circuit boards mounted vertically on either side of the camera motor.

If the camera fails to function correctly, and the possibility of battery failure has been eliminated, all circuit boards can swiftly be changed in the field by the Camera Assistant.

A spare set of circuit boards (and a motor cover tool for the GII etc.), is supplied with all cameras. Experience has shown, however, that the likelihood of an electronic circuit failure is very small.

Before changing the circuit boards, the battery-to-camera cable should be replaced and the camera tested on a spare battery pack to ensure that the power supply to the camera is adequate.

BEFORE CHANGING ANY ELECTRONIC BOARD THE CAMERA MUST BE COMPLETELY DISCONNECTED FROM THE BATTERY.

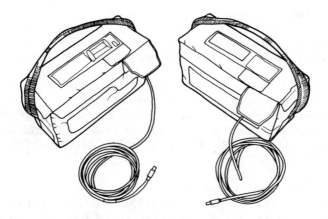

Camera battery and cable + spare battery and cable. BEFORE ALL ELSE TRY
USING A SPARE BATTERY AND A SPARE CABLE.

When changing boards it must be all or nothing.

Changing the Electronic Circuit Boards

A spare set of circuit boards is supplied with every PANAFLEX camera. PLATINUM PANAFLEXES have three circuit boards, the others all have four.

To gain access to the motor compartment where the circuit boards are located it is necessary to remove the motor cover.

Before removing the motor cover the power supply to the camera must first be disconnected and the battery cable removed. THE POWER SOURCE MUST BE DISCONNECTED BEFORE CIRCUIT BOARDS ARE CHANGED. THIS IS *VERY* IMPORTANT.

Removing the motor cover
The motor cover of a PLATINUM PANAFLEX may be removed by undoing the screws in each corner of the motor cover and removing it gently.

The motor covers of the GII and earlier PANAFLEXES are released by turning two locks which are accessed from inside the camera.

To unlock the motor cover, push the motor cover tool through two openings in the soundproof lining. One is directly below the pitch control knob in the bottom left-hand corner of the film compartment, the other is above the buckle switch on the right-hand side. To release, turn the locks counter-clockwise as far as they will go.

To remove the motor cover lift it slightly and then pull it out from the underside.

Changing the circuit boards
Release the spring-loaded retainers from the circuit boards and remove ALL boards by holding by top and bottom and pulling straight out. Replace ALL boards.

Note: The edge connectors of all boards are different so that no board can be plugged into the wrong camera or the wrong position.

Replace the motor cover and tighten securely.

In the event that a set of boards are changed the complete set of discarded boards should be returned to PANAVISION Inc. or their Distributor as speedily as possible accompanied by a note detailing the circumstances and symptoms that necessitated the change. PANAVISION Inc. or its Distributor will quickly supply a second set.

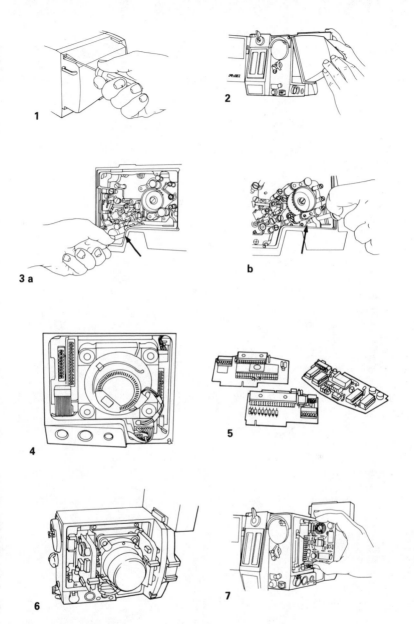

1. PLATINUM PANAFLEX motor cover release screws, 2. Removing a PANAFLEX motor cover, 3. GII motor cover locks access holes: a. below the pitch control knob, b. above the buckle switch, 4. Interior of PLATINUM PANAFLEX showing position of circuit boards, 5. Set of PLATINUM PANAFLEX circuit boards, 6. Interior of GII showing position of circuit boards, 7. Changing a circuit board.

167

Using
PANAFLEX and PANASTAR
Camera Accessories

Bringing the power of the computer to the aid of the cinematographer.

Cinematographers' Computer Calculator

The PANAVISION CINEMATOGRAPHERS' COMPUTER CALCULATOR comprises the CINEMATOGRAPHERS' COMPUTER PROGRAM already installed into a Sharp PC-1360 pocket computer.

It instantly calculates all the many variables encountered in cinematography.

While Depth of Field and Focus Split calculations are obviously the most frequently used sub-programs, others which calculate the safe operating parameters for HMI lighting, etc., are equally useful.

The programs may be run in Feet and Inches or in Metric units. Once changed to a particular system the programs will remain that way until changed again.

Summary of programs

ACCOUNTS: Given salary rates, time worked and any other payments calculates Total Salary.

DIMENSIONS: Converts Meters to Feet and Inches, Feet and Inches to Meters and calculates Triangulated Distances.

DIOPTERS: Calculates Nearest and Furthest Distances with a diopter and Required Diopter.

EXPOSURE: Calculates Exposure Adjustments, Exposure Times, Key Light-Fill Light ratios, Contrast Ratios, etc.

FILM TIME/LENGTH: Calculates Film Length/Running Times, Total Length and Aggregate Length in Feet and Frames or Meters and Frames, etc.

FOCUS: Calculates Depth of Field, Focus Split/Minimum Aperture, Focus Split/Maximum Focal Length, Focus Split/Near and Far Distances, Hyperfocal Distance, etc.

FORMATS: Gives sizes of approx. 30 different Film, TV and Still formats.

SAFE HMI SETTINGS: Calculates safe Camera Speeds, Shutter Angles and Supply Frequencies.

LIGHT and ELECTRICITY: Calculates and converts various Lighting and Electrical parameters.

MACRO: Calculates Lens Extension distances, Magnification factors, corrected Exposure Settings, overall Depth of Field, etc.

OPTICS: Calculates Object/Screen distances, Lens Focal Length/Angles, Scene Widths and Heights, Screen and Image sizes, etc.

SPEEDS: Calculates Miniature/Scale speeds, Screen and Object speeds, Speeded-up and Slow Motion speeds, etc.

SUN TIMES: Calculates Dawn, Sunrise, Sunset and Dusk times for any location on any date.

170

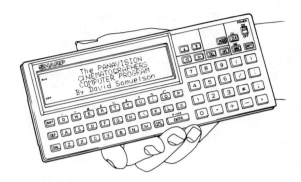

Sharp PC-1360 micro-computer containing the PANAVISION CINEMATOGRA-
PHERS' COMPUTER CALCULATOR

Individual programs may be accessed via a Main Menu (press the key marked DEF
followed by the key marked SPC) or directly by pressing the DEF key followed by
the Program Letter:

Keys	Program	Easy way to remember if first letter is not available
DEF + A	ACCOUNTS	
DEF + B	PANAVISION LOGO	B for beginning of program
DEF + C	COLOR	
DEF + D	DIOPTERS	
DEF + F	FOCUS	
DEF + G	FORMATS	G for gauges
DEF + H	HMI	
DEF + J	SUN TIMES	J for Jour, French for day!
DEF + K	FILM TIME/LENGTH	K for Kodak film
DEF + L	LIGHT & ELECTRICITY	
DEF + M	MACRO	
DEF + N	DIMENSIONS	N for Numbers
DEF + S	SPEEDS	
DEF + X	EXPOSURE	
DEF + Z	OPTICS	Z for Zeiss
DEF + =	CHANGE NUMERIC SYSTEM	Ft & Ins = Meters
DEF + SPC	MAIN MENU	Next to the ENTER key

Electronic Synchronizer Systems

The sync speeds of PANAFLEX cameras are derived from quartz crystal oscillators to accuracies better than one frame in one hour ensuring safe synchronization with sound recorders and HMI lighting, providing they are speed controlled to a similar degree of accuracy. Even if the speeds were to drift to their maximum during the course of the longest possible take (11 minutes, the running time of a 1000 ft. roll of film) there would still be no noticeable loss of sync or discernible flicker. (Note: Some early-model PANAFLEX cameras, which have not yet been returned to the factory for updating, may still be fitted with a crystal which is only accurate to within one frame in 20 minutes.)

In the case of the PLATINUM PANAFLEX the camera can be crystal controlled at any speed, in 0.1 fps increments, from 4 to 36 fps, including 24, 25, 29.97 and 30 fps.

Earlier PANAFLEXES usually have crystal-controlled speeds of 24 and 25 fps but can be supplied with 24/30 or 24/29.97/30 fps on request. Cameras fitted with the latest circuit boards can be switched to any of the four speeds.

There are occasions, however, when the camera must be positively synchronized with other equipment (HMI lighting, video systems or computers), or must be run in shutter phase sync with a projector (front or back projection), or with another camera (twin camera 3D), or it may be desirable to run the camera at speeds outside the normal camera speed range (time lapse, single shot), etc. In all these instances it is necessary to override the normal speed control system and sync to an outside source.

Electronic synchronizing accessories, which are totally automatic in their operation, are available for all PANAFLEX cameras. These accessories also function with PANASTAR, SUPER PSR and 65mm cameras.

172

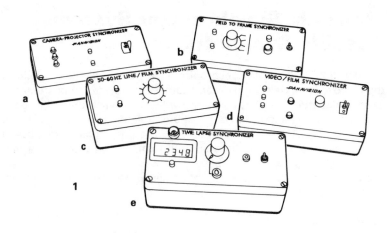

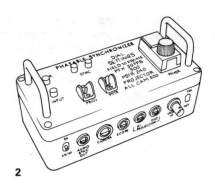

1. Electronic control boxes available for use with PANAFLEX cameras:
a. Camera-Projector Synchronizer, b. Field to Frame Synchronizer, c. 50-60 Hz
Line/Film Synchronizer, d. Video/Film Synchronizer, e. Time-lapse Synchronizer,
2. Phasable Synchronizer unit.

173

Electronic Synchronizer Systems— Phasable Synchronizer

The PANAVISION Phasable Synchronizer incorporates all the capabilities of the Video to Film, Field to Frame and Camera to Projector Synchronizers.

Its special feature is a phasing control which can synchronize the speed of the camera and the position of the shutter relative to an external pulse over a range of $\pm 100°$ ($\pm 180°$ on request). The phasing is accomplished while the camera is running in a smooth manner and with only a momentary 1% change in camera speed.

The phase knob has a "turns" counting dial and a precision 10-turn pot for accurate resetting.

Operating instructions

Phasing can be accomplished while filming and observing the effect through the viewfinder:

While filming a TV monitor set the camera shutter to 180° and turn the phase control knob until the bar is out of sight (Field to Frame); or set the shutter to 144° and position the bar to be at its least objectionable (Video to Film).

For Front Projection set the camera shutter to its full open position and turn the phase control knob until the least amount of light is seen through the viewfinder. The maximum amount of light is then passing to the film.

174

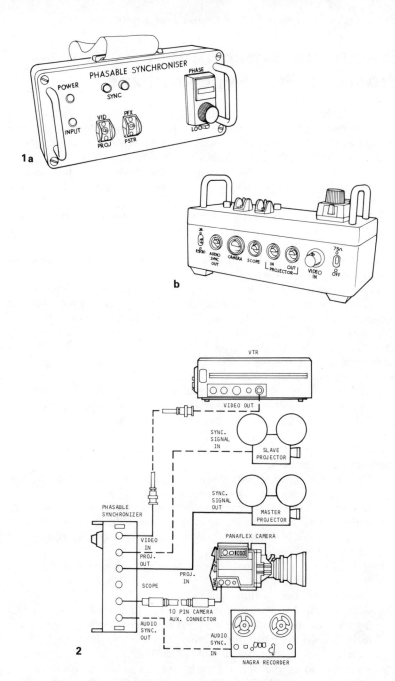

1.a. & b. Phasable Synchronizer control box, 2. Phasable Synchronizer schematic of connections.

Synchronizes a 24 fps camera with a 60Hz video display.

Electronic Synchronizer Systems— Video to Film Synchronizer

The Video to Film Synchronizer will eliminate the drifting bars effect from the TV monitor that appears within a scene. *This synchronizer can only be used with a 60 Hz NTSC TV system and a PANAFLEX camera incorporating a 10-PIN, LEMO-type accessory output connector.*

The factors that must be taken into consideration are the film camera speed of 24 frames per second and the video field rate of 60 fields per second.

When the shutter is correctly adjusted to the monitor display, the synchronizer will properly phase the film camera to the video and allow the video system to control the camera speed. The camera will photograph the monitor screen with a very fine line in the center of the monitor every other frame.

When projected, this line is not evident and should not be objectionable.

Absolute synchronization may be verified by removing the film and replacing the aperture plate with an Aperture Viewing Mirror.

The Aperture Viewing Mirror may also be used to view a video display exactly as it will be seen by the film and at the same time adjusting the shutter opening, by means of the micrometer shutter adjuster, until the bar line is most narrow.

The Audio Sync connector makes available a 1 volt peak to peak 60 Hz signal for audio synchronization. It may be desirable to use this on a long scene with the sound since the video system may not hold the film camera at exactly 24 frames per second.

Operating instructions

1. Connect the synchronizer to the video source. If not otherwise terminated, set the termination switch to 75Ω.
2. Connect the synchronizer to the film camera with the 10-pin LEMO cable supplied with the kit. The 10-pin connector on the Camera is located under a small gray plastic hatch. On the synchronizer the connector is labelled CAMERA.
3. If required, connect the synchronizer to the audio recorder.
4. Set the camera shutter to 144°.
5. Switch the Video to Film Synchronizer to VIDEO.
6. Start the camera. The red Search Indicator LED will flash for the first 3 to 5 seconds of the synchronized operation while the camera is locking onto the video signal, after which the green Sync LED will light, indicating that the camera is locked onto the video signal.
7. To return to normal filming speed set the switch at the top of the synchronizer to CAMERA DRIVE.

176

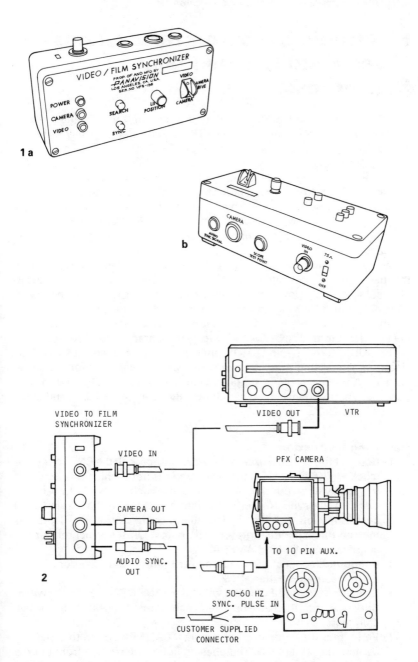

1.a. & b. Video to Film Synchronizer control box, 2. Video to Film Synchronizer schematic of connections.

177

Electronic Synchronizer Systems— Video Field to Film Frame Synchronizer

The PANAVISION Video Field to Frame Synchronizer unit enables any PANAFLEX camera (except the PANASTAR) to be electronically locked to almost any video source, including NTSC, CCIR (PAL & SECAM) and 48 FIELD (24 frame).

Filming with a video-to-film frame ratio of 1:1 completely eliminates the rolling bar effect from the image of the video screen by locking the film camera to the video scan rate. This means that with a 59.94 Hz NTSC type video signal THE CAMERA WILL RUN AT 29.97 FRAMES PER SECOND. Equally, it will run at 25 fps with a 50 Hz CCIR video input and at 24 fps with a 48 FIELD input.

The system will also work with Video Tape Recorders and Computers that have either Composite Video, Composite Sync., or Vertical Sync available as an output. However, in these situations it may be necessary to use an Aperture Viewing Mirror to precisely adjust the camera shutter to set the exposure period to match that of the video source. This is a task best carried out by PANAVISION or its representatives worldwide.

The AUDIO SYNC (3-pin LEMO) connector makes available a 1 volt RMS sinewave signal that may be used as a pilotone reference for sound recording. A mating cable is supplied with bare wires on one end for this purpose. The frequencies are as follows: 60 Hz at 24/24, 30/30 and 30/24 and 50 Hz at 25/25.

Operating instructions

1. Connect the synchronizer to the video source with the BNC connector labelled VIDEO INPUT. If the video signal is not terminated elsewhere in your system, put the black recessed switch to the 75Ω position.
2. Connect the synchronizer to the film camera with the 10-pin LEMO cable supplied with the kit. The 10-pin connector on the Camera is located under a small gray plastic hatch. On the synchronizer the connector is labelled CAMERA.
3. If required, connect the synchronizer to the audio recorder.
4. Set the camera shutter at 180°.
5. Select the Video-to-Film ratio with the FRAMES PER SECOND switch.
6. Supply 24v to the camera. The synchronizer POWER indicator will illuminate.
7. Verify that the video is on and switch CAMERA DRIVE to VIDEO.
8. Switch the camera ON. The phasing lights will change from the red to GREEN when phasing is complete. Looking through the viewfinder there should be no sign of a horizontal line moving up or down.

178

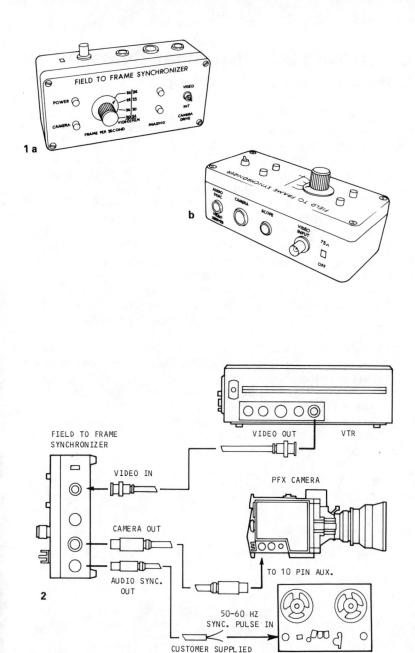

1.a. & b. Field to Frame Synchronizer unit, 2. Field to Frame Synchronizer schematic of connections.

Electronic Synchronizer Systems— Line to Film Synchronizer

The PANAVISION 50-60 Hz Line to Film Synchronizer unit enables the frames per second rate of any PANAFLEX, PANASTAR or SUPER PSR camera to be locked to a 50 or 60 Hz power supply. At the same time it produces a sync pulse which can be fed to the sound recorder.

With a 60 Hz power supply any camera may be run at 24 fps. PANASTAR cameras may additionally be run at 60 or 120 fps.

With a 50 Hz power supply any camera may be run at 25 fps and PAN-ASTAR cameras may additionally be run at 50 or 100 fps.

Line to Film Synchronizers are mostly used to lock the camera speed to the same AC power source that is supplying the HMI or other short arc discharge light source illuminating the set. It is particularly useful when the lighting is being powered by a portable alternator (generator) which is not fitted with a crystal controlled frequency speed governor.

It is the most effective way to eliminate any possibility of flicker due to the camera and the lighting running at incompatible speeds.

Operating instructions

1. Plug the Line/Film Synchronizer into the lighting mains supply to take a reference frequency.
2. Connect the Line/Film Synchronizer to the camera via the 10-pin Auxiliary connector.
3. If required, connect the Audio Sync outlet to the Audio recorder.
4. Set the rotary switch on the top of the Line/Film Synchronizer to the correct Hz/fps setting.

180

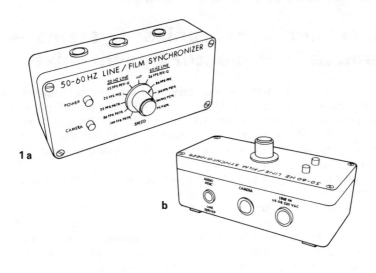

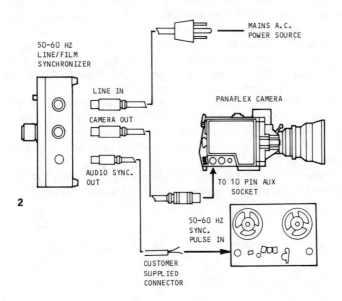

1.a. & b. Line to Film Synchronizer unit, 2. Line to Film Synchronizer schematic of connections.

Electronic Synchronizer Systems— Camera to Projector Synchronizer

The PANAVISION Projector Electronic Synchronizer unit not only ensures that any PANAFLEX, PANASTAR or SUPER PSR camera will operate at exactly the same speed as a film projector but will also ensure (providing that the probe is properly set on the projector) that the shutters of the two units open at precisely the same moment. It is an essential accessory for front and rear projection process filming.

To operate the Camera to Projector Synchronizer unit an electronic probe unit must be fitted in close proximity to a raised stud attached to some part of the projector which rotates once every frame. This will generate a reference pulse which will replace the internal crystal control pulse within the camera and to which the camera will automatically synchronize. A suitable probe unit for the projector can be supplied on request.

The stud and the probe unit must be set relative to one another so that they pass at the moment when the projector shutter is 48° past its closed position. It is advantageous if the probe unit can be rotated relative to the stud to allow for fine adjustment both of synchronization and of the relative exposure between the foreground and background of the process shot. If the stud can be rotated 180° relative to its normal shooting position this setting can be used during rehearsals to check the lighting balance between the projected image and the foreground.

The clearance between the stud and the probe should be approximately ¹⁄₁₆″, 1.5mm. If it is too close double pulsing may prevent sync; if there is too much clearance no pulse will be generated. If necessary adjust the probe in or out until the camera synchronizes properly.

The AUDIO SYNC (3-pin LEMO) connector makes available a 1 volt RMS sinewave signal that may be used as a pilotone reference for sound recording. A mating cable is supplied with bare wires on one end for this purpose. The frequencies are as follows: 60 Hz at 24/24, 30/30 and 30/24 and 50 Hz at 25/25.

Absolute synchronization may be verified by removing the film and replacing the aperture plate with an Aperture Viewing Mirror.

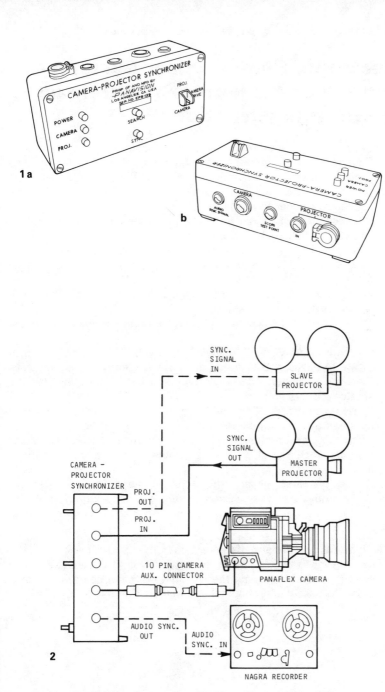

1.a. & b. Camera to Projector Synchronizer unit, 2. Camera to Projector Synchronizer schematic of connections.

183

Electronic Synchronizer Systems— Camera to Projector Synchronizer Operating Instructions

1. Set the camera shutter to its fully open position.
2. Connect the synchronizer to the film camera with the 10-pin LEMO cable supplied with the kit. The 10-pin connector on the Camera is located under a small gray plastic hatch. On the synchronizer the connector is labelled CAMERA.
3. Connect the sync cable from the projector shutter probe to the PROJECTOR IN socket on the rear of the synchronizer. (The PRO-JECTOR OUT connector is used when two or more cameras are synced to the same projector when a sync box is required for each camera.)
4. If required, connect the audio pulse cable to the AUDIO SYNC SIG-NAL socket and to the recorder.
5. The POWER light will come on when the battery is connected to the camera.
6. Switch the CAMERA DRIVE switch to PROJ. (When this switch is in CAMERA position, the camera runs independently of the projector.)
7. Start the projector. The PROJ. light will flash, indicating that the shutter pulse is present.
8. Start the camera. The CAMERA light will flash. Initially the red SEARCH light will also light, indicating that the camera is not yet at speed. After a few seconds the red light will go out and the green SYNC light will flash indicating that the camera is now up to speed and is in sync.
9. As a safety check, observe the projected image through the mirror shutter reflex viewfinder system of the camera. Its brightness should be minimal, indicating that the maximum amount of light is being used to expose the film.

Notes: It makes no difference if the camera or the projector is started first.

It is important to check that the green SYNC light continues to flash steadily during a take and that the red light does not come on. Should this not be the case it is probable that a malfunction has occurred.

184

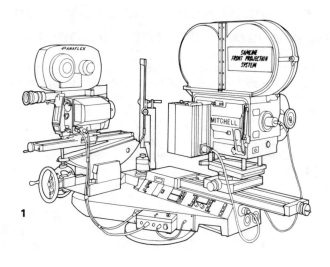

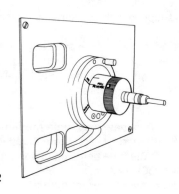

1. Front projection rig using a PANAFLEX camera and showing the projector synchronizer box in the foreground, 2. Rotatable probe unit fitted to a Front Projection projector.

Making events look faster than they really are.

Electronic Synchronizer Systems— Time-Lapse and Time Exposures

A Time-Lapse/Time Exposure speed control accessory is available which makes it possible to run PANAFLEX and PANASTAR cameras at speeds slower than the 4 fps, which is the minimum speed with the normal camera speed control.

In the TIME-LAPSE mode all exposures are made as if the camera were operating at a constant rate of four frames per second. The exposure time may be controlled by using the adjustable shutter.

In the TIME-EXPOSURE mode the duration of each exposure depends on the SECONDS PER FRAME setting. Exposures longer than 640 sec. (10⅔ min.) can be obtained by using the SINGLE FRAME button.

Note: As a precaution it is advisable to black out the magazine and camera housing with duvetine or similar material when very long time lapse sequences are envisaged or in very bright ambient light conditions.

Operating instructions

1. Select either TIME-LAPSE or TIME-EXPOSURE operation with the CAMERA DRIVE switch.
2. Connect the time-lapse synchronizer to the film camera with the 10-pin LEMO cable supplied with the kit. The 10-pin connector on the camera is located under a small gray plastic hatch. On the box the connector is labelled CAMERA.
3. When power is supplied to the camera, the green POWER light should illuminate.
4. In the TIME-LAPSE mode set the switch marked X1 and X10 before selecting the frame rate. In the X10 position all of the seconds per frame settings (except .25) are multiplied by TEN.
5. Select the frame rate as required.
6. Set the camera shutter opening to control exposure.

The CAMERA light will illuminate each time the film advances.

In the TIME-EXPOSURE mode it is recommended that the shutter opening be set at 180°. At 180° the time between exposures (the film pull-down period) is fixed at 1/18 (.056) sec.

Note: A special motor cover, incorporating a blower unit, must be used with the PANASTAR camera when scenes or re-takes last for 30 minutes or longer. For long running times with a PANAFLEX camera, a special magazine that has reduced power is required.

186

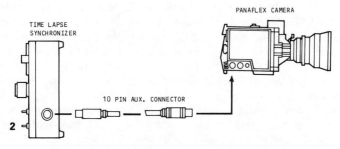

1. Time lapse control box, 2. Time lapse schematic of connections.

TIME-LAPSE EXPOSURE TABLE

Shutter opening (°)										
200	180	160	144	120	100	80	70	60	50	40
Exposure time (sec)										
$\frac{1}{7}$	$\frac{1}{8}$	$\frac{1}{9}$	$\frac{1}{10}$	$\frac{1}{12}$	$\frac{1}{14}$	$\frac{1}{18}$	$\frac{1}{21}$	$\frac{1}{24}$	$\frac{1}{29}$	$\frac{1}{36}$
0.14	0.125	0.11	0.1	0.08	0.07	0.06	0.05	0.04	0.35	0.28

EVENT TIME/SCREEN TIME CALCULATIONS

If the elapsed time of the event and the desired screen time are known the necessary frame rate may be calculated viz.:

Seconds per frame = Event time / (Screen time × 24)

If the elapsed time of the event and the frame rate are known the screen time may be calculated viz.:

Screen time = (Event time × 24) / Seconds per frame setting

All times must be in seconds.

187

Enhancing the shadows without affecting the highlights.

PANAFLASHER

Giving a very small, even, overall exposure to the film immediately before or after the film is exposed in the camera gate is a method of increasing effective film speed, reducing contrast, putting more information into the shadow areas and, if desired in the case of color film, of tinting the shadows without affecting the highlights. It is a technique which is as old as photography itself.

The PANAVISION PANAFLASHER unit fits onto whichever PANAFLEX or PANASTAR magazine port is not currently in use. It makes no difference if the film is flashed just before or just after the principal exposure.

The effect of flashing is most pronounced in the shadow areas. It has little or no effect in the middle tones or the white areas of the picture. In consequence, scenes require less fill light and deep shadows will show more detail than they might otherwise have done.

If the flashing light is colored it will color the dark areas only. It may be used to give an overall warming or cooling effect. Moonlight, sepia and virtually any other mood may be created by using inexpensive gelatin and/or 49mm diameter screw-in filters in the PANAFLASHER.

The amount of flashing may be adjusted during the course of a scene and is particularly effective when colored flashing is being used. The effect of increasing the flashing during a take will first be seen in the darkest areas and will spread to the highlights as the PANAFLASHER iris is opened until the entire scene can have an overall colored look about it.

General guidelines

Before using the PANAFLASHER system it is important that carefully controlled photographic tests be carried out to determine what the overall effect will be in the circumstances of a particular lighting situation.

When the scene is contrasty and low-key, about 10% flash exposure is likely to be a good basis for initial tests. In this situation comparatively little flash exposure is required because the shadow area comprises most of the image.

When a scene is contrasty but high-key, 20% flashing can be used to bring out the shadow details.

When a scene is high contrast, with equal areas of light and shade, 15%-20% flashing should be tried.

188

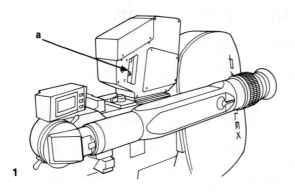

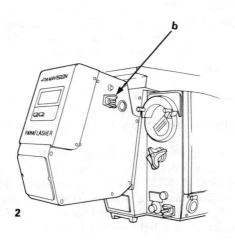

1. PANAFLASHER unit mounted on the top magazine port; a. PANACLEAR power supply socket and switch, 2. PANAFLASHER unit mounted in rear position; b. on-off switch.

189

The amount of light used in flashing is very small so it is important to get it correct.

PANAFLASHER—
Operating Instructions

1. Put the appropriate filters in the filter tray:

NEUTRAL	As specified by PANAVISION
WARM	" " " " + 85B
COOL	" " " " + 80A
COLOR	As required and determined by tests.

2. Attach the PANAFLASHER unit to whichever magazine port is not in use.
3. If desired, plug the PANACLEAR power supply cable into the socket provided on the PANAFLASHER unit.
4. Switch ON the PANAFLASHER unit. (A green LED indicates ON.)
5. If the PANAFLASHER is mounted on the *rear* magazine port push the EI/TIME button to the EI position and set the EI to equal the Exposure index of the film in use, using the up and down arrows next to the button, i.e. 5247 = EI 125.
6. If the PANAFLASHER is mounted on the *top* magazine port push the EI/TIME button to the EI position and set the EI according to the TOP MOUNT table opposite, i.e. 5247 = EI 40. (Note: It is necessary to make this adjustment to compensate for the fact that the PANAFLASHER unit is farther from the film in the top position than at the rear.)
7. If running the camera at any speed other than 24/25 fps reset the EI according to the CAMERA SPEED table opposite.
8. Push the EI/TIME button to the TIME position and set the speed to $\frac{1}{50}$ sec.
9. Push in the M-CLR button and move the IRIS control back and forth to set the EV number to give the desired amount of flashing (see BASIC FLASHING INTENSITY table opposite.) Note: This table is only supplied as an example. Please refer to the table supplied with the individual PANAFLASHER unit for the actual E.V. settings.
10. Check the EV reading before every take.

Note: The PANAFLASHER consumes very little power and it is not necessary to switch it OFF between takes.

190

PANAFLASHER INTENSITY CHART	
%	E.V. number
5%	3.2
10%	3.4
15%	3.7
20%	3.9
25%	4.2
30%	4.7
0%	4.9

TOP MOUNT PANAFLASHER EI COMPENSATION		
EI	becomes	EI
500	"	160
400	"	125
320	"	100
250	"	80
200	"	64
160	"	50
125	"	40
100	"	32
80	"	25

CAMERA SPEED CHART								
EI for the film in use								
24 fps	500	400	320	250	200	160	125	100	80
BECOMES								
6 fps	2000	1600	1200	1000	800	640	500	400	320
12 fps	1000	800	640	500	400	320	250	200	160
18 fps	640	500	400	320	250	200	160	125	100
30 fps	400	320	250	200	160	125	100	80	64
48 fps	250	200	160	125	100	80	64	50	40
96 fps	125	100	80	64	50	40	32	25	20

PANAGLIDE Floating Camera Systems

A PANAGLIDE floating camera outfit consists of a vest (which the Operator wears and which supports the entire system), a stabilizer arm (which smooths out bumpy movements), an upright arm (which supports the camera at one end and the control unit/counterweight at the other and which may be mounted upside down for low-angle camera positioning), a control unit/battery container, special 24v batteries plus spares, a battery charger system, a lightweight video camera and monitor, a remote Focus and T stop control unit and a docking tripod (to support the rig between takes).

The cameras most frequently used with PANAGLIDE floating camera systems are special lightweight versions of the PANAFLEX-X, which combine ease of use with silence of operation. Other possible choices are a PANAFLEX 16 or a lightweight PANAVISION 65mm camera.

Additional available accessories include the means to attach a PANAGLIDE suspension arm to a vehicle or a tripod head.

The lightweight PANAFLEX-X camera may be fitted with a pellicle reflex mirror in place of a spinning reflex mirror to give a flicker-free video image at the cost of 1/3 stop in exposure.

PANAGLIDE type lightweight cameras are available separately to use with Steadicam floating camera systems.

The PANAGLIDE vest and suspension arm

The PANAGLIDE vest may be put on the operator like a jacket and secured by safety-belt type buckles.

The suspension arm attachment plate may be attached to the vest either way round to mount the camera to the left or the right of the operator, to suit either the left or right-handedness of the operator or to better accommodate a particular shot.

A quick-release ring is fitted to the vest by the operator's right shoulder to enable him to quickly divest himself of the rig in an emergency. This is particularly important when working close to a swimming pool or other deep water.

192

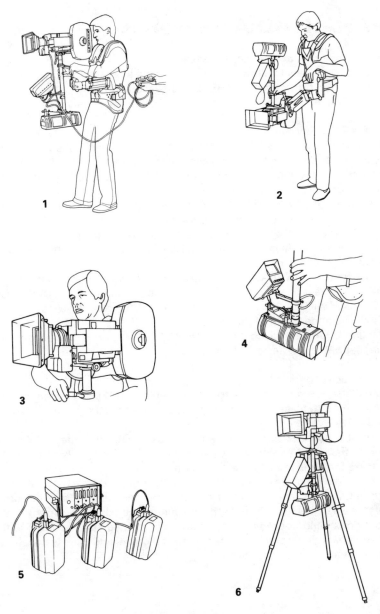

1. PANAGLIDE floating camera in normal mode, 2. PANAGLIDE floating camera in inverted mode, 3. Lightweight PANAFLEX-X camera, 4. Control unit/battery container and video monitor mounted in low position, 5. PANAVID multiple battery charging system, 6. PANAGLIDE camera system "at rest" on a docking tripod.

193

PANAHEAD Accessories

Accessory box attachment plates

Attachment plates to allow Camera Assistants to attach their own "front" boxes to the front of a PANAHEAD are available from the store at PAN-AVISION (the "PANASTORE") and from their representatives worldwide.

PANABALL leveller

An optional accessory for the PANAHEAD geared head is the PANABALL Leveller. This may be fitted between the tripod head and the PANAHEAD or used as a levelling hi-hat and stood, bolted or screwed directly to the ground or any flat surface.

To attach the PANAHEAD to the leveller tighten up the topmost of the two spoked handwheels set in the leveller.

To level or tilt the PANAHEAD release the ball movement by turning the lower spoked handwheel in a clockwise direction. Adjust the level and re-tighten the lower handwheel.

The regular PANABALL leveller gives 15° of levelling and tilting movement. A 3° version is available to use in conjunction with a PANAVISION NODAL ADAPTOR which enables a PANAHEAD to be levelled or tilted about the nodal point of the camera lens.

If the flange onto which the PANABALL leveller is to be fitted is too wide the feet of the leveller may be removed.

Nodal adaptor

Nodal adaptors are available to fit between a PANAFLEX camera and the PANAHEAD to place the entrance pupil of the lens in the center of the pan and tilt axes of the head.

To fit a nodal adaptor remove the tilt assembly from the PANAHEAD and fit the nodal adaptor in its place, using the slide facility to place the camera and lens in the correct position.

To set the entrance pupil of the lens on the nodal point of the head place two pointers in line in front of the camera, one close to and the other farther away. Looking through the viewfinder the relative positions of the pointers should not move as the camera is panned and tilted. Move the camera backwards and forwards and up and down until there is no relative movement.

194

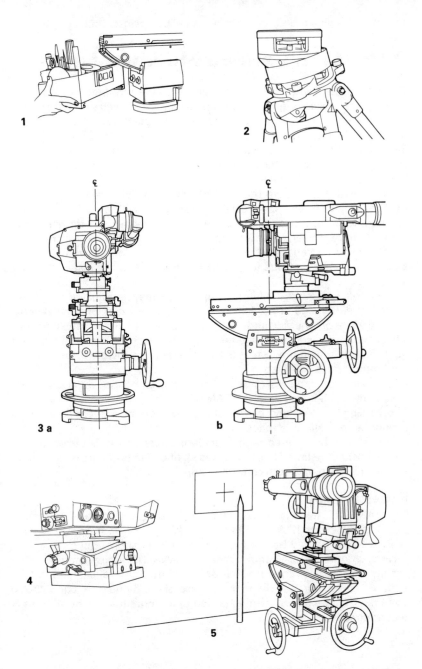

1. PANAHEAD front box attachment plate, 2. 15° PANABALL leveller on a tripod, 3.a. & b. Camera set up nodally with a 3° PANABALL leveller and a nodal adaptor set on the ground, 4. Nodal adaptor, 5. Nodal setting up method.

195

The adjustable brightness/constant color temperature camera light.

PANALITE Camera Mounted Lamp

The PANALITE Camera Mounted Lamp may be mounted close to the camera lens and its brightness controlled without affecting the color quality of the light. It is mostly used to fill facial shadows. The adjustable brightness facility, which works by varying the area of the reflecting surface, enables the light intensity falling on the subject to remain constant as the distance between the camera and the subject is reduced or increased during the course of a take.

The PANALITE fits onto the support bracket which is slipped over the iris rods (matte box support bars). It is supplied with an offset arm and adapters to make it possible to place the lamp higher and farther to the left or right of the lens than normal. A ball joint on the rear of the lamp housing allows it to be tilted in any direction. It is supplied with two power cables (one short and one long) fitted with in-line ON-OFF switches and a spare bulb.

In the soft light mode the intensity of the light may be adjusted by turning a knob at either side of the lamp which rotates a series of circular rods, painted white on one side and black on the other, which form the reflector. When all the white sides of the reflector tubes face the bulb the reflected light is at its maximum, when the black sides are towards the bulb the minimum light is reflected.

The extension cable used for the follow focus and iris and shutter controls may also be used with the PANALITE.

Reversing the bulb and its reflector gives a bright incident light which cannot be intensity controlled.

The PANALITE is fitted with four barn doors and is supplied with a complement of gelatin filter frames which fit on support bars which swing out from the top and bottom of the lamp.

PANALITE bulbs
The PANALITE takes a $4^{11}/_{16}$", 117mm, linear quartz bulb. The recommended bulbs are: 120v × 650 watt, 220-230v × 875w and 240-250 volts × 875w. They should be used with an AC power supply only.

Note: In order to prevent overheating PANALITE lamps should not be left switched on continuously but only lit during the rehearsal and take periods. This is particularly so in the case of the higher wattage 220-250 volt bulbs, where a maximum lit period of no more than eight minutes is recommended.

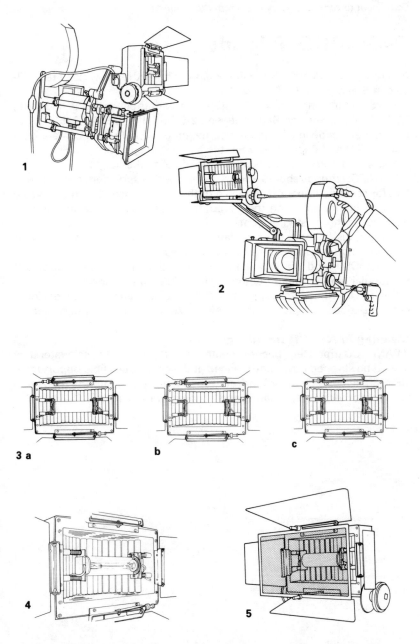

1. Normal PANALITE mounting, 2. Offset PANALITE mounting using a follow focus extension for remote control, 3. PANALITE with a. an all white reflector, b. a 50/50 reflector, and c. an all black reflector, 4. PANALITE with lamp turned towards the subject, 5. PANALITE with barn doors and gelatin filter holder fitted.

PANAPOD Tripods

PANAVISION PANAPODS are lightweight tubular tripods which are available in Standard and Baby lengths.

The Standard length makes possible lens heights between 3'9" and 6'10", (1.150-2.80 m.) and the Baby between 2'9" and 4'3" (0.840-1.295 m.), approximately, allowing for the added height of a PANAFLEX camera on a PANAHEAD and depending upon the amount of spread.

The use of a PANABALL leveller will add 6" to the height.

PANAPOD tripods should always be used with a spreader or with crows' feet for security. If the feet are fixed to the ground they should always be released before adjusting the tripod height or level to prevent the legs from becoming twisted or bent.

Three rings beneath the top plate, central with each leg, may be used for attaching tie-down chains or safety ropes.

PANAPOD tripod legs should never be overtightened as this may distort the tubular legs, making them difficult to slide up and down.

PANAPOD tripods are supplied with a shoulder pad to make it easier to carry the tripod complete with a PANAHEAD and camera, if necessary.

Cleaning PANAPOD tripod legs

If PANAPOD tripod legs become contaminated with sand or saltwater they should be stripped down and cleaned at the earliest possible opportunity. NEVER use oil to clean or lubricate tripod legs. It gets into the metal and will dirty other peoples' hands and clothing from then on.

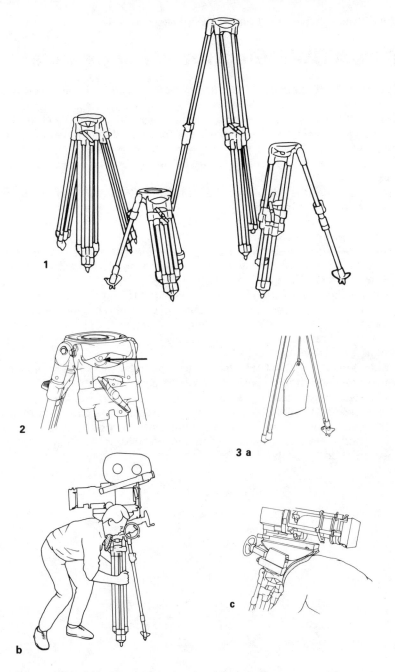

1. Standard and baby PANAPOD tripods at their shortest and most extended heights, 2. PANAPOD tripod top showing tie-down eye, 3.a., b, & c. Using a shoulder pad to carry a camera, head and tripod assembly

PANATAPE Electronic Rangefinder

The PANATAPE electronic rangefinder measures the distance from the film plane to a selected solid object which is (usually) placed centrally in front of the camera. It is intended primarily as a system to confirm distances which have previously been eye-focussed or made with a tape measure, rather than as a principal means of measurement. It is particularly useful when the camera is tracking in or out on a subject, or vice versa.

In most cases the PANATAPE has a range of 2-15' (0.6-4.5m) with an accuracy of ±1" (25mm).

Fitting and operation

The PANATAPE detects distances by means of two ultrasonic sensors which are mounted above the matte box by means of a bracket which fits on the iris rods (matte box support bars).

Lateral and vertical adjustments are provided to align the sensors with any particular foreground object although, in general, it is better to confine the use of the PANATAPE to a single central object.

The digital read-out is mounted onto a wide-angle matte box which is supplied with the system.

Before use, the PANATAPE electronic rangefinder should be set up and tested at a number of distances to verify its accuracy.

Any necessary adjustments may be made by the control knob on the side of the display unit. Occasionally, the PANATAPE may pick up spurious signals which may affect its accuracy but these errors are usually quite gross, and therefore obvious, and thus can easily be detected and discounted or compensated for.

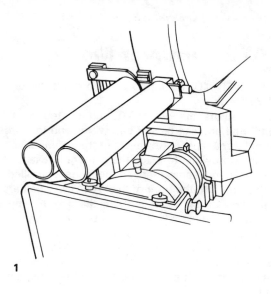

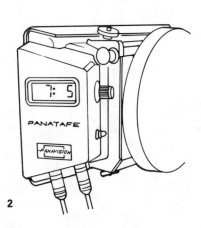

1. PANATAPE sensors on top of camera, 2. PANATAPE read-out on matte box.

The means to make the world turn upside down.

PANATATE Turnover Mount

PANATATE TURNOVER MOUNTS are available to tilt the camera about the lens axis during a shot.

Like the Nodal Adaptors, the PANATATE unit is fitted to a PANAHEAD in place of the tilt unit and can be slid backwards and forwards to place the entrance pupil of the lens in the center of the pan and tilt axes.

Fine adjustment is provided to place the camera in exactly the center of rotation so there is no displacement as the camera is rolled over.

Weights which screw into the front of the PANATATE unit are provided to partially counterweight the camera about the pan, tilt and rotational axes so that it may be panned, tilted and rotated smoothly.

Stretch the battery and any other camera cables out from the rear of the camera so they do not wrap around the PANATATE unit as it is rolled, and unwind between takes.

The Camera Operator may find it easier to operate using a video viewfinder during a take.

The PANATATE unit takes a normal PANAHEAD handwheel and has exactly the same speed change system. It may also be used with the PANAMOTE remote PANAHEAD control system.

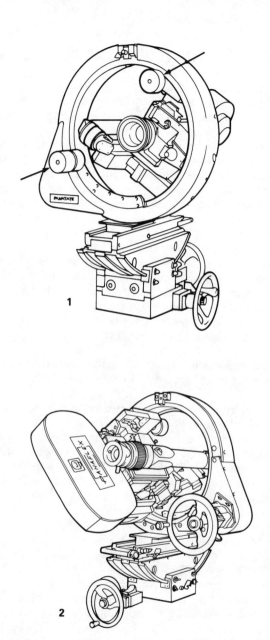

1. PANATATE turnover mount, front view showing weights, 2. PANATATE turnover mount, rear view showing handwheels.

203

A marriage of film and video cameras.

PANAVID Video Assist Systems

Since lightweight, simple-to-operate, video assist systems were first introduced PANAVISION has been at the forefront of video assist development, always advancing the state of the art and only using off-the-shelf equipment when it is advantageous to do so.

The current models include the state-of-the-art CCD PANAVID and PANAFRAME system, which offers flicker-free video images at any filming speed, the capability of being locked to the frame rate of the film camera, the facility of grabbing a single frame for analysis or for subsequently overlaying another image and the ability to use the infra-red part of the spectrum to gain additional sensitivity in low-light conditions.

Another current model is the SUPER PANAVID, which incorporates the most sensitive Newvicon video tube available.

All PANAVID video assist systems incorporate a HIGH GAIN circuit to enable the sensitivity to be boosted by the equivalent of two Stops of light (12 db). This is particularly useful when high-speed filmstocks, high EI ratings and forced film development are in use.

They also incorporate a neutral density filter which may be inserted into the lightpath to compensate for bright sunlight conditions and slow filmstocks.

Back Light Compensator circuits enable the contrast ratios to be adjusted to favor the more important action in either the shadow or the highlight areas.

Both systems may be remotely controlled and/or externally driven by outside horizontal and vertical drive signals.

The PANAFRAME system incorporates an electronic framestore circuit which eliminates video flicker caused by the rotating mirror reflex shutter on the film camera. The SUPER PANAVID can be made flicker free by using the PANAVISION Video Flicker Processor or by replacing the rotating mirror reflex shutter with a pellicle reflex mirror at the cost of ⅓-½ Stop of exposure to the film.

PANAVID accessories

The following accessories are available to use with PANAVID systems:

VIDEO FLICKER PROCESSOR: For use with the SUPER PANAVID. Removes flicker from video at 24 and 30 fps.

IMAGE ENHANCER: Rings the video image to improve definition.

FRAME LINE GENERATOR: Superimposes black or white frame lines on the video. The positions of the frame lines are fully adjustable.

ON-BOARD MONITORS: Small monitors which may be mounted on the film camera for the benefit of the Operator and/or the Focus Assistant.

204

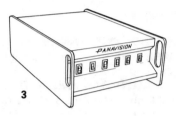

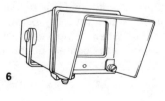

1. CCD PANAVID camera unit, 2. PANAFRAME Mk.I control unit,
3. PANAFRAME Mk.II control unit, 4. SUPER PANAVID camera unit, 5. Video
FLICKER PROCESSOR control unit, 6. Small video monitor for on-board usage.

205

PANAVID—CCD PANAVID
Operating Instructions

When ordered with a PANAFLEX camera, a PANAVID video assist unit will already have been fitted to the film camera before it is delivered to the user. However, there may be a need for a Camera Assistant to remove it in the field in order to gain access to the motor compartment to change the circuit boards.

To remove a PANAVID camera unit first remove the retaining screw at the back of the unit (the side facing the viewfinder) using an alien key. Carefully slide the unit rearwards to disengage it from the electronic connectors.

PANAVID—CCD PANAVID controls
POWER switch: Controls power to the CCD camera independently of the film camera power.

IRIS AUTO/MAN switch: The AUTO position provides automatic control of both the lens iris and electronic gain of the video camera, responding more to large areas of image brightness than to small areas of high brightness or deep shadow. It may be switched to MAN to manually set the video sensitivity to favor a particular area of image brightness. When changing between AUTO and MAN the existing sensitivity setting is retained as the initial setting and the correct ratio between IRIS and GAIN are maintained in either mode.

AUX Connector: Provision for attaching future accessories.

VIDEO OUT: A 75Ω BNC connector for connecting the CCD PANAVID camera to the camera input socket of the PANAFRAME unit. NOTE: THE VIDEO OUTPUT OF THE CCD PANAVID IS NOT SUITABLE FOR CONNECTING DIRECTLY TO A NORMAL VIDEO MONITOR OR RECORDER AND MUST PASS THROUGH THE PANAFRAME UNIT.

OPTICAL FILTER WHEEL: May be set to "N" for normal lighting conditions (full visible spectrum but no infrared light), "ND" for bright ambient lighting conditions (12% of visible light only) and "IR" to use full visible plus infrared light for maximum video sensitivity in poor lighting conditions, or when using very fast filmstocks, or underexposing and force developing the camera negative.

206

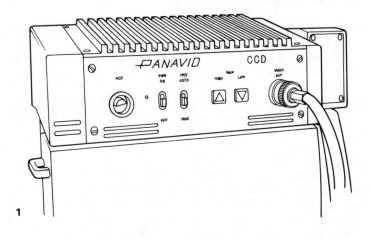

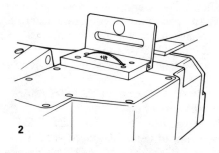

1. CCD PANAVID camera unit, 2. CCD PANAVID filter wheel.

A box of black and white magic.

PANAVID—PANAFRAME Operating Instructions

The PANAFRAME unit is complementary to the CCD PANAVID video assist camera. The one cannot be used without the other.

The PANAFRAME unit is an electronic image processing system which now enables the video signals from the CCD unit on the film camera to be passed flicker-free to normal video monitors and recorders. PANA-FRAME also offers facilities to freeze a single picture frame for close examination or to allow it to be used as an "electronic matte frame" to be matched to a live image.

PANAFRAME front panel controls
POWER switch: Switches between OFF, 24 volt battery operation or an AC power supply. The AC supply may be any voltage between 90 and 240 v and 50 or 60 Hz. A red light indicates when battery power is in operation and green indicates AC power. Both may be connected at the same time and the switch used to select whichever is required.

MODE switch: Switches between NORM (normal real time flicker-free video assist images), FREZ (a frame of picture from either the video assist camera, or from a recorder, or other video source is captured and frozen on screen) and COMP (frozen and live images are alternated on the screen 30 times a second so that they may be compared).

Notes: In the NORM position the video assist image will automatically synchronize with the film camera and will remain flicker-free down to a camera speed of 4 fps. An image may be captured and frozen at any time but the film must be stopped when a live image from the film camera is compared with a frozen image (COMP mode).

EXT switch: Switches image input between the video assist camera and any other external source.

IRIS switch: A remote control for the IRIS AUTO/MAN switch on the CCD camera unit with the same effect.

DETAIL switch: Controls the amount of detail visible in the image. In the OFF position the amount of detail displayed is reduced, in the MAX position it is overemphasized. The NORM position is standard.

SYNC lamps: Indicate if the internal timing circuits are locked to either the video assist camera or to an external source connected to the GEN-LOCK input on the rear panel.

CAMERA "ON" lamp: Indicates when the film camera is running.

208

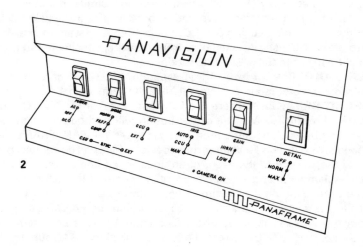

1

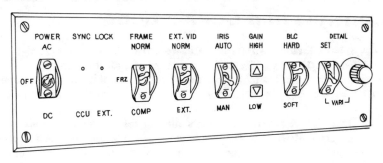

2

1. PANAFRAME Mk.I front panel, 2. PANAFRAME Mk.II front panel.

209

There are two sides to every good thing.

PANAVID—PANAFRAME Rear Panel Controls and Connections

PANAFRAME rear panel sockets and controls

CAMERA VIDEO IN: 75Ω BNC connector used for input from the video assist camera only. Up to 100 meters of 75Ω co-axial cable may be used.

EXT VIDEO IN: 75Ω BNC connector for input from any standard video source.

75 Ohm TERM switch: Connects internal 75Ω terminating resistor for EXT VIDEO IN. Must be ON unless video is being looped through to other units, in which case the video signal must be terminated at its far end.

GEN LOCK connector: Used either to lock two (or more) PANAFRAME units together or to lock the PANAFRAME unit to a stable source of standard video. The signal may be sync only or composite video. Being high impedance, an external terminating resistor may be desirable.

HARD/SOFT CLIP switch: The HARD position is normal. In the SOFT CLIP position the video image contrast is reduced by compressing the brightness range in the highlight areas of the picture, leaving the shadow areas untouched. The SOFT CLIP position acts in a similar manner to the shoulder of a film transfer curve and is most useful in heavily backlit situations. Note: When using SOFT CLIP it is often necessary to use MAN IRIS to prevent the strong highlights from reducing sensitivity too much as might happen in the AUTO IRIS mode.

VIDEO OUT Connectors: Three separate 75Ω BNC connectors for video monitors and recorders, etc. All have the same signal.

CIRCUIT BREAKERS: There is a 2 Ampere breaker for 24v DC and a 1 Ampere breaker for AC. The center button will pop out when tripped and is reset by pushing the button back in. Note: The PANAFRAME also has an internal Ground Fault Interrupter device which will disconnect both sides of the AC line in the event of an unbalanced line current of more than 5 milliamperes and for this reason it is important to ensure that the ground connection on the AC supply line is always good. Should the GFI device trip it may indicate an internal fault within the PANAFRAME unit and for this reason the cause of the fault should always be determined before resetting the GFI by momentarily disconnecting the AC supply. The GFI does not operate on the 24v DC supply.

POWER connectors: There are separate AC and DC connectors. 24v DC is through a 2-pin Lemo connector with the negative terminal of the battery grounded to the unit case. The AC supply is through a 3-pin connector and cord with the center pin grounded to the unit case. An interference filter is included as part of the AC connector.

210

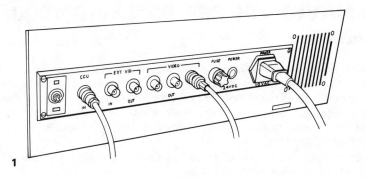

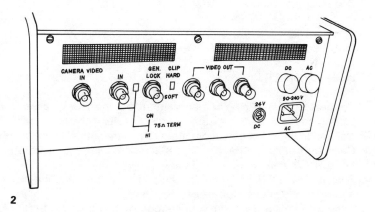

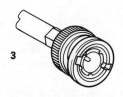

1. PANAFRAME Mk.I rear controls and connections, 2. PANAFRAME Mk.II rear controls and connections, 3. 75Ω BNC connector.

211

PANAVID—SUPER PANAVID 1000 System

For normal usage a SUPER PANAVID 1000 camera unit may be connected directly to a video monitor or recorder and a power supply and on-board controls used to adjust the picture image quality settings. Alternatively it may be operated remotely using a SUPER PANAVID CONTROL UNIT or made flicker free, and operated remotely, by using a SUPER PANAVID FLICKER PROCESSING unit.

Both of these units may be used in conjunction with a Genlock sync lock system. (See the following page.)

SUPER PANAVID operating instructions

1. Before connecting the SUPER PANAVID to a power supply, check that the switches are all set as follows:

   Power—OFF
   BLC (Back Light Compensator)—SOFT
   Gain—NORMAL
   Iris—AUTOMATIC

2. Connect SUPER PANAVID camera unit to a monitor or a video recorder/playback unit using BNC coaxial cable. If the monitor is the last unit in the chain of video units set the termination switch to 75Ω.
3. Connect the SUPER PANAVID to a 24 volt power supply using the Sony 10-pin connector to XLR cable.
4. Switch power ON and wait for the picture to appear. The iris will automatically correct for proper exposure.
5. If the video picture is not bright enough, switch the Gain to HIGH to inject a 12db (2 stop) gain boost.
6. Set the Back Light Compensator to hard or soft depending upon the scene lighting.
7. If shooting outside in full sunlight or using a very slow filmstock, when the automatic iris may not be able to stop down the video camera sufficiently to control the video brightness, use the Neutral Density switch on the camera head to reduce the light to the video by two Stops.
8. An LED on the SUPER PANAVID unit will flash when the power supply falls below 22 volts. The battery should be changed and the old battery recharged.

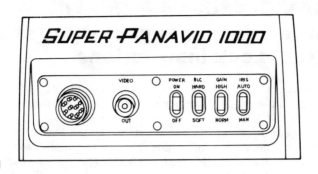

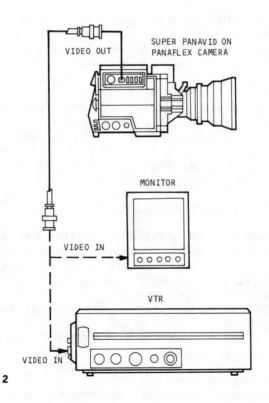

1. SUPER PANAVID camera, 2. SUPER PANAVID connection directly to a monitor or video recorder schematic.

PANAVID—SUPER PANAVID
Remote Control, Flicker Processor
and Gen-Lock Instructions

The functions of a SUPER PANAVID unit may be controlled remotely and/
or it may be connected to a Genlock sync system by the use of a SUPER
PANAVID CONTROL UNIT.

Video images from a normal video assist unit normally suffer from flicker
while the film camera is running due to the intermittent nature of the
mirror shutter reflex viewfinder system. This problem may be eliminated,
and the SUPER PANAVID unit operated remotely and connected to a Gen-
lock system, by the use of a SUPER PANAVID FLICKER PROCESSOR.

SUPER PANAVID CONTROL UNIT operating instructions

1. Attach the SUPER PANAVID camera unit to the film camera and set
   the switches as described on the previous page.
2. Connect the SUPER PANAVID camera unit to the CONTROL UNIT
   using the cable fitted with a 10-pin Sony connector.
3. Connect the CONTROL UNIT to a 24v DC or to an AC power source
   as required and switch on. Note: The unit will automatically detect
   if the power supply is AC or DC and react accordingly. Units supplied
   in the U.S. will normally be wired for a 110v AC power supply and
   those elsewhere for 220/240v. If a unit is being taken to another
   country where the power supply may be different, then either PAN-
   AVISION or its representatives will supply a unit for the required
   voltage.
4. Proceed as in steps 6-9 on page 212.

SUPER PANAVID FLICKER PROCESSOR operating instructions

1. Attach the SUPER PANAVID camera unit to the film camera, set the
   switches, connect the SUPER PANAVID camera unit to the FLICKER
   PROCESSOR unit and connect the FLICKER PROCESSOR unit to a
   power source as described above.
2. Connect the video monitors and/or the video recorder to the FLICKER
   PROCESSOR unit.
3. Set the camera speed switch on the FLICKER PROCESSOR unit to
   match the camera speed.

Gen-lock operating instructions

When using an external drive (Gen-lock) with either the SUPER PANAVID
CONTROL UNIT or the FLICKER PROCESSOR the external drive switch
inside the PANAVID unit must first be set to the *rear* position. To gain
access it is first necessary to remove the top cover of the PANAVID camera
unit.

214

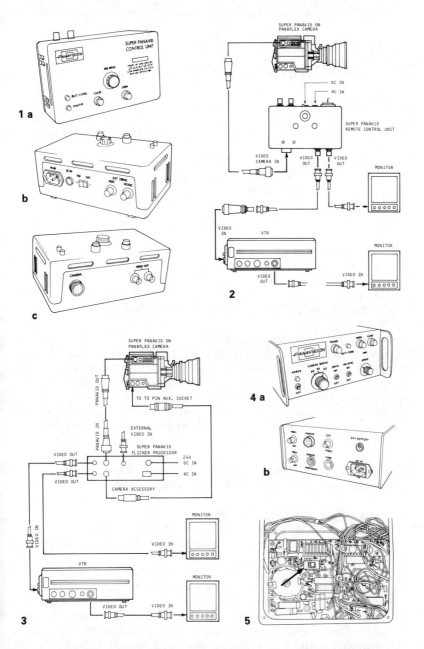

1.a., b. & c. SUPER PANAVID remote control unit, 2. SUPER PANAVID remote control schematic, 3. PANAVID FLICKER PROCESSOR schematic, 4.a. & b. PANAVID FLICKER PROCESSOR, 5. "External sync" switch inside a SUPER PANAVID head.

Time Code—
The AATONCODE System

The AATONCODE time code system, developed by AATON, the French camera manufacturer, is a means of encoding every frame of film with the SMPTE Time Code in a manner that can be read both by a computer *and* a human being.

The computer-readable information may be used by a telecine machine fitted with an AATON TIME CODE READER or by an appropriately equipped film editing table to automatically synchronize picture to sound.

The secret of the AATONCODE time code marking on film lies in the method of exposing a matrix of 7 × 13 dots along the edge of the film to register both the time code and to create the normally readable alphanumeric characters.

The AATONCODE Time Code system comprises three units: the GENERATOR (fitted to the film camera), the portable ORIGIN C (used to transport the master time from the recorder to the camera) and the NAGRA IV-S T.C. (fitted with an AATON serial board modification).

Initializing the time code system
At the beginning of each shooting day the Camera Assistant must initialize the system by checking, and if necessary, entering the time, the date and the relevant production information into the system and synchronizing the camera and the Nagra together.

The individual clocks in both the camera and the recorder will then remain in absolute sync for at least four hours, after which they need to be briefly connected together again in order to maintain absolute sync for another four hours.

The equipment must be re-synchronized if the camera has been disconnected from a power source for 45 minutes or more.

Encoding the film during the take
Whenever the camera is running the AATONCODE system will *automatically* mark the exact time and every frame number during a second on every frame of film and on the audio tape.

Other information, such as Production number, the camera I.D. letter, etc., will be marked onto the film once every second.

A slate displaying the name of the Production and the Production Company should still be photographed onto the head of every roll of film and recorded onto every roll of tape as a means of identification.

For telecine transfer it is recommended that a traditional slate be used briefly to visually identify each scene and take, and to use the time code as a means to automatically synchronize dailies quickly, accurately and effortlessly.

216

The AATONCODE system

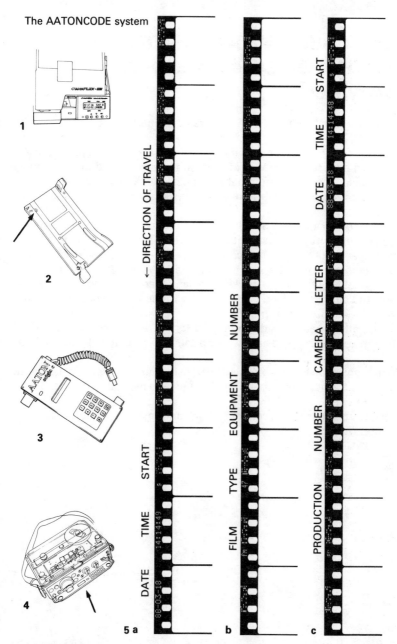

1. AATONCODE Generator module fitted to a PANAFLEX camera, 2. PANAFLEX aperture plate (3 PERF) showing AATONCODE exposure opening, 3. AATON ORIGIN C module, 4. Time code panel on a Nagra IV-S T.C. recorder, 5. a., b. & c. Strip of film showing time code edge display sequence.

217

Time Code—PANAVISION AATONCODE Generator Unit Overview

The PANAVISION AATONCODE Generator is fitted to the underside of a PANAFLEX camera.

The Generator, together with the special camera aperture plate and the unit which exposes the matrix of dots on the edge of the film, will have been installed into the camera before it goes onto a particular production. The Camera Assistant will not have any installation procedures to worry about.

The Generator takes its power from the film camera (it has no ON-OFF switch) and is programmed with the camera serial number before the camera leaves PANAVISION Inc.

The Generator must be synchronized with the Time Code unit within the Nagra at the start of the working day and thereafter at least every four hours, and after any period when power has been disconnected from the camera for 45 minutes or more.

If the battery has been disconnected for more than 45 minutes a red LED light will flash and a message on the LCD display will say *OUT OF SYNC.*

Synchronization is accomplished by transferring the time from the Nagra IV-S T.C. to the more portable ORIGIN C unit and using that to set the time in the camera.

It is good practice, when convenient, to confirm that the units are holding absolute sync.

218

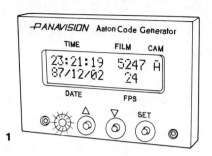

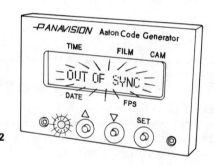

1. Generator panel displaying Time Code data settings, 2. Generator panel displaying the *OUT OF SYNC* message and the red LED flashing.

Time Code—PANAVISION AATONCODE Generator Unit Data Entry

The PANAVISION AATONCODE Generator will retain its real time and date settings even when the camera has been disconnected from a power source. The rest of the data need only be entered when something changes.

Entering and setting data into the AATONCODE Generator unit
Data must be entered into the Generator unit in the following sequence:

1. Set the **FILM TYPE.**

The Film Type controls the exposure of the time code "dots" on the film (and for this reason it is imperative that it be reset correctly every time the filmstock is changed, even for a single take).

The current Film Type setting ("5247", for instance) will always be displayed.

Press the SET button. The Film Type in the display will flash, prompting the user to set it.

Press the up and down keys to scroll backwards and forwards between the various filmstock types.

When the correct film type is displayed press the SET button.

2. Set the **CAMERA LETTER.**

After the Film Type has been set, the Camera Letter will flash.

To change the Camera Letter again press the up and down keys to scroll backwards and forwards through the alphabet.

Press the SET button when the appropriate letter is reached.

3. Set the **FPS RATE.**

After the Camera Letter has been set the FPS will flash.

As before, this is done by the use of the up and down keys and by pressing the SET button when the appropriate frame rate is displayed.

Note: At this time the system is capable of handling 24, 25, 29.97 non-drop and 30 fps time code rates. If the camera is being under or over cranked for effect the fps rate should be left at the normal production sync sound speed.

4. Set the **PRODUCTION NUMBER.**

After the fps rate has been set, the last production number entered will flash. However, the production number setting may be skipped since this information will be loaded from the Nagra (via the Origin C unit).

Note: When any part of the display is flashing, waiting for an input, for longer than 30 seconds, and nothing is touched, the display will stop flashing. The SET button must then be pressed again to change the data.

220

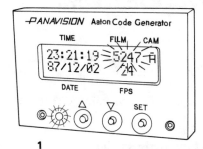

1

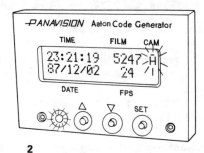

2

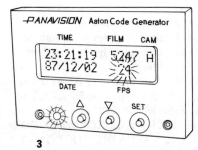

3

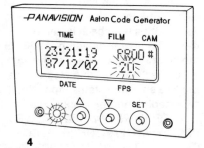

4

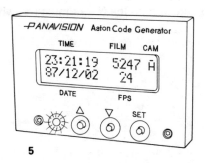

5

1. Generator panel with *SET FILM TYPE* display flashing, 2. Generator panel with *SET CAMERA LETTER* display flashing, 3. Generator panel with *SET CAMERA FPS* display flashing, 4. Generator panel with *SET PRODUCTION #* display flashing, 5. Generator panel with red malfunction LED flashing (call PANAVISION, or its Representative, if this happens).

Time Code—AATONCODE/Nagra IV-S T.C. Unit

The recommended audio recorder for Time Code operation is the stereo Nagra IV-S model fitted with a Time Code module, making it a "Nagra IV-S T.C."

In addition the Nagra recorder must be equipped with an AATONCODE ASCII interface board.

Checking and changing the Nagra time code settings

To operate correctly in the time code mode the frame rate must be set to that of the film camera, i.e. 24, 25 or 30 fps. This may be done by means of a rotary switch situated inside the recorder. Note: This need not necessarily be the same as the camera frame rate. At 24 fps on the camera 24 or 30 fps may be selected on the Nagra depending upon the nature of the audio post production. At 25, 29.97 non-drop or 30 fps on the camera the Nagra should be set at the same frame rate.

To check the existing Nagra time code settings press the **STATUS** key once to enter the status mode and then press the adjoining **NEXT ST** key repeatedly until the display reads either *FrEE Ub* or *dAtE Ub*.

NOTE: IF *FrEE Ub* IS DISPLAYED, THE NAGRA WILL NOT WORK CORRECTLY WITH AATONCODE. CONSULT THE NAGRA INSTRUCTIONS TO CHANGE TO *dAtE Ub* MODE BEFORE INITIALIZING THE NAGRA FOR AATONCODE.

Before initializing the system the Nagra must be switched to **TEST**.

222

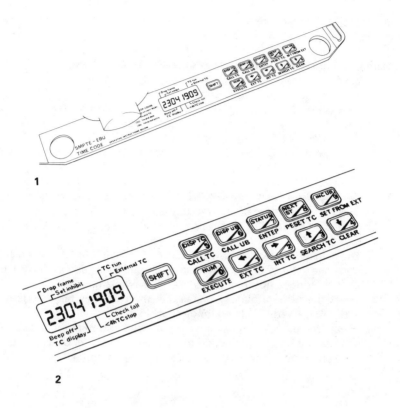

1. Nagra time code control panel, 2. Nagra time code key pad.

Time Code—Initializing the System

In practice the master time is normally taken from the time code clock inside the Nagra IV-S T.C. recorder.

Transferring data from the Nagra to the ORIGIN C unit
1. Verify that the Nagra is in *dAtE Ub* mode (see page 222).
2. Put the Nagra into *TEST* mode.
3. Switch the ORIGIN C unit ON, plug it into the Nagra recorder and press the "*" key. The display should show *In Control* then *Good.* If not press the "*" key again.
4. Disconnect the ORIGIN C unit from the Nagra but *do not switch it OFF.*

Transferring data from the ORIGIN C unit to the camera
1. Check that power is connected to the camera.
2. Plug the ORIGIN C into the camera unit socket and press the "*" key.
3. The display will read *bAd. . . . REloAd?.* While *REloAd* is displayed press the "*" key again. The display should then read *In Control* or *Good.* If it does not then press the "*" key again.
4. Check the camera display to see that the code has been acquired and that the red LED is no longer flashing.
5. Repeat the process for each camera making sure that the ORIGIN C unit has not been switched off meanwhile. Note: The ORIGIN C unit will automatically switch itself off after three minutes.

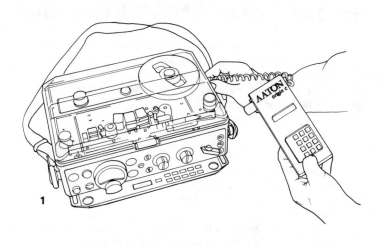

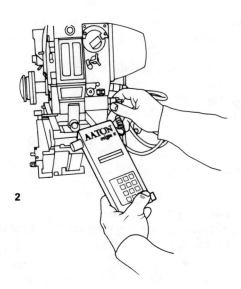

1. Transferring data between a Nagra and an ORIGIN C unit, 2. Transferring data between an ORIGIN C unit and the AATONCODE Generator on a PANAFLEX camera.

225

The AATONCODE go-between.

Time Code—AATON ORIGIN C Unit

The AATON ORIGIN C unit is used to carry the exact time from the time code unit in the Nagra IV-S T.C. recorder to the Generator unit in the camera.

The ORIGIN C unit may also be used to reset the date and the time settings in the Nagra, if required.

Inputting data
Original data can be entered into the ORIGIN C unit as follows:

1. Switch the unit ON.
2. The display will first ask for the Production Number (PR).
3. Enter a two digit number, i.e. "45."
4. Press the "#" key.
5. The display will then ask for the year (Y), followed by the month, the day, the hour, the minutes and the seconds.
6. Enter two digits into each of them and press the "#" key after each entry.
7. Press the "*" key to start the clock running.
8. Press the "#" key and hold down to see the clock running.

Resetting the Nagra time and date settings
To reset the Nagra time and date settings from the ORIGIN C unit (which should already have been set correctly) plug it into the 5 pin Lemo socket on the right-hand side of the Nagra and then press the "*" key. The ORIGIN C display should read *bAd. . . . REloAd*

While REloAd is displayed again press the "*" key on the ORIGIN C. The display will say *In Control. . . . Good.*

The ORIGIN C unit may then be disconnected from the Nagra and used to set the camera.

Periodically checking Time Code sync
Every two or three hours, or when convenient, the Time Code sync should be checked to be sure it is running and good:

1. Switch the unit ON and check that power is connected to the camera.
2. Plug the ORIGIN C unit into the Nagra, switch the Nagra to TEST and press the "*" key. The recorder time will be acquired by the ORIGIN C and the display will say *GOOD.*
3. Plug the ORIGIN C cable into the camera unit and press the "*" key. The display should also read *GOOD.* If it reads *FAIR* or *BAD* reload the time code into the camera by pressing the "*" again.

226

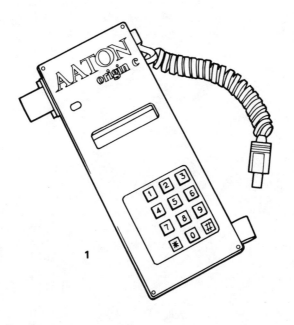

1. The AATON ORIGIN C unit.

227

The
Sound Recordists'
PANAFLEX

Creating the Quietest Possible Recording Conditions

From the Sound Recordist's point of view, the most important aspect of the choice of camera for a particular production is "how much noise is the principal camera going to make when shooting close in to the leading artist who is speaking in whispers in absolutely quiet surroundings?".

With a PANAFLEX camera the chances are that the noisiest thing around will be the Nagra recorder.

Vast amounts of time, money, energy and ingenuity have been expended by PANAVISION to develop and refine the quiet running aspect of the PANAFLEX range of cameras in order that the maximum percentage of original sound is usable and does not have to be looped because there is unacceptable camera noise in the background.

This has been done without compromise to the weight and the size of the camera, the way the lenses are mounted, the steadiness of the image or the size of the registration pins and the usability of the camera in general.

The PLATINUM PANAFLEX, in particular, requires no extra blimping for even the closest microphone positioning.

Just as the quietness of the camera has been achieved by a scientific approach and a team effort at the factory, so the maintenance of that state of quietness requires good servicing, a knowledgeable use of the equipment in the field and the cooperation of all the crew.

Measuring camera noise

All cameras make a noise, especially with film running through them in the intermittent manner that film runs through a camera, and this is measured over a wide range of frequencies.

PANAVISION measures the noise level of its cameras with a full roll of film running through the camera, with the microphone placed 3' (0.914m) from the film plane, with the magazine in the top position so that it is not hidden from the microphone, with any lens on and in an exceptionally quiet sound-proof chamber.

This test is significantly more exacting than making measurements with the microphone 1 meter (3'3.37") away and using only a short focal length lens.

PANAVISION also make polar graphs of the camera noise so they can check the amount of noise emerging from the camera in all directions (because the microphone is rarely placed directly in front of the lens) and measure the spectrum of camera noise, from which they can determine the amount of noise being generated by every individual moving component within the camera.

230

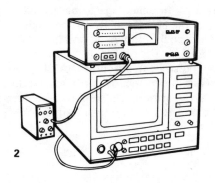

1. PANAFLEX camera in noise test room, 2. Noise data recording instrument.

Camera Noise Control

If a camera becomes noisier than it should then there are a number of possible causes that can be checked immediately, viz.:

Is the filter door closed?
Is the film running over the sprocket teeth correctly?
Are both loops as large as possible?
Has the pitch control been set for optimum quietness?
Is the de-anamorphoser lever set properly?
Are the aperture plate locking levers touching the camera door?
Is anything creating a mechanical link between the mechanism plate and the camera body?
Are all the internal circuit boards secure?
Is the motor touching the motor cover?
Does the movement require lubricating?

If none of the above is the answer to the problem then the camera should be returned to PANAVISION, or its representative, for attention.

Minimizing camera noise

Although the noise generated by PANAFLEX cameras is reasonably even all around, there is a slight advantage to be gained by positioning the microphone in front of the film plane (rather than behind the camera) and at "10 o'clock" or "5 o'clock" relative to, and looking at, the front of the camera.

Most of what little noise comes from PANAFLEX cameras is confined to a few very narrow frequency bands. Much can be done, particularly at the dubbing stage, to minimize any recorded camera noise by putting in notch filters at 24 Hz and 12 KHz and their harmonics.

232

1

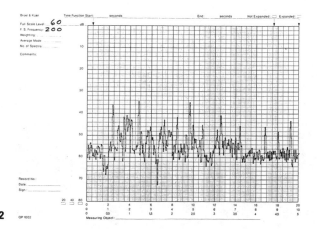

Brüel & Kjær   Time Function Start:____ seconds   End:____ seconds   Not Expanded: ☐   Expanded: ☐

Full Scale Level: **60**   dB
F. S. Frequency: **200**
Weighting:
Average Mode:
No. of Spectra:
Comments:

Record No.:
Date:
Sign:

QP 1002   Measuring Object:

2

3

1. Polar camera noise graph showing noise level three feet forward from the film plane, 2. Noise spectrum graph of a PLATINUM PANAFLEX, 3. Noise spectrum graph of a GII PANAFLEX.

233

The
Production Managers'
PANAFLEX

Ordering PANAVISION Equipment

To the Production Manager falls the task of ordering the Camera and associated equipment, and of making sure that what is needed is where it has to be, when it has to be. Equally he or she must make sure that whatever is no longer needed is returned as soon as possible, so that money is not spent renting equipment unnecessarily.

To make the operation run as efficiently as possible PANAVISION and its Representatives worldwide have specially trained client liaison personnel who are never more than a telephone call, or a fax, or a telex away. Equally, all of PANAVISION's equipment is only a truck or an aircraft journey away.

At the start of a production it is always advantageous to try to have a personal meeting with the person who is going to look after your project, so that when you make telephone calls later on you both know who you are talking to.

Whenever practical PANAVISION's client contact person will appreciate being invited to visit the set or the location to meet and to talk to the people who are actually using the equipment. It not only helps to make the current production run more smoothly but also affords an opportunity to hear comments and suggestions which can be incorporated into PANAVISION's product range in the future.

Ordering over the telephone

When you place an order over the telephone PANAVISION's client contact person will write it down on a special form he has for the purpose. It is not necessary but it does make life easier, and there is less likely to be an error, if you state your requirements in the same order as is printed on the form. To this end PANAVISION will always supply you with copies of their order form so that everyone is using the same document.

PANAVISION Taken By _____ Order No. _____ Prod. I.D. _____

Producer: _____ Ordered by: _____ Date: _____

Picture: _____ Telephone: _____

Ship: _____ Start: _____ Ret: _____ P.O.: _____ Prod.: _____

Deliver to: _____ Asst. Prep.: Yes ☐ No ☐ ___ Advance ☐ Ins ☐

Location, Etc.: _____ Revision 1 2 3 4 Will Comp. D.P.: _____

Address: _____ P.M. _____

Item	Lens A		Lens B		Lens C		Lens D	
PLATINUM 2.35 1.85 T.V.	16	T2.8	20	T3/T4	24	T1.6	30	T3
PANAFLEX 2.35 1.85 T.V.	24	T2.8	28	T2.8	35	T2.3	40	T2.8
PFX-X 2.35 1.85 T.V.	32	T2	35	T2	50	T2.3	55	MAC.
	40	T2	50	T2	75	T2.3	100	T2.8
FLICKER FREE R.C.U.	75	T2	100	T2	150	T3.5	180	T3
SUPER-PANAVID C.C.U.	400	T4	500	T4	360	T4	600	CAN
	1000	T6.3	40	MAC.	800	T5.6	1000	T5.6
1000 PLAT MAGS. (REV)	60	MAC.	90	MAC.				
500 PLAT MAGS.					35	T1.4	50	T1.4
1000' MAGS. HTR. CVR.	24	T2	28	T1.9	50	T1.1	75	T1.8
500' MAGS. HTR. CVR.	35	T1.6	50	T1.4	100	T1.8	85	T2
250' MAGS. HTR. CVR.	55	T1.1	300	T2.8				
AUX. CARRYING HANDLE/VID	300	NIKON	300	CAN	28	T2	35	T1.4
PFX ZOOM CONT. HOLDER	600	CAN	400	CAN	40	T1.4	50	T1
SUPER FOLLOW FOCUS	800	CAN						
					40/200	T4.5		
PFX UTILITY BASE	14	T1.9	17	T1.9 (C.F.)	50/500	T5.6		
PANALENS LIGHT UNIT	20	T1.9	24	T1.2	PANAHEAD			
RAIN COVER	29	T1.2	35	T1.3	O'CONNOR HEAD			
WEATHER PROT LG./SH.	40	T1.3	50	T1.0	SACHTLER HEAD			
REMOTE SWITCH	75	T1.6	100	T1.6	STANDARD (G) (B)			
SUNSHADE EXTENSION	125	T1.5	150	T1.5	BABY (G) (B)			
PANALITE	200	T2	PRIMO		SPREADER			
AUTO BASE PFX/PFX-X/VIDEO	"Z" SERIES		UZ SERIES		HI HAT (MITCHELL)			
PFX MID-RANGE EYEPIECE					SLATE			
PANASTAR 2.35 1.85 T.V.	20/100	T3.1			CHANGING BAG			
SUPER-PANAVID C.C.U.	20/120	T3			FLANGE FOCAL DEPTH SET			
1000' MAGS. (REV.)	25/250	T4 (SZ)			PANAHEAD ADAPTER PLATE			
500' MAGS.	23/460	T10			SPARE ZOOM CONTROL (N) (O)			
ARRI IIC / ARRI III 2.35 1.85 T.V.	2X EXTENDER				FOAM EYEPIECE COVERS			
CEI VIDEO	MATTE BOX MATTES				EYEPIECE LEVELER (N) (O)			
NORMAL SPEED/HIGH SPEED	6 × 6 MATTE BOX				PANATILT			
CRYSTAL MOTOR	24V. BLOCK COMP.				INCLINING PRISM			
CONSTANT SPEED	24V. BELT COMP.							
VARIABLE SPEED	24V. PURSE COMP.				PROCESS SYNC BOX			
HIGH SPEED + REO. + CABLES	24V. L.A. BATT.				FIELD/FRAME SYNC BOX			
400' MAGS.	36V. COMP.				VIDEO/FILM SYNC BOX			
200' MAGS.					144° BRKT & MIRROR			
ARRI BASE PLATE								
ARRI PERISCOPE	85, N3, N6, N9				T H	4×5	6×6	138 4.5
ARRI HI HAT ADAPTER	LC 1, 2, 3, 4, 5.				T H	4×5	6×6	138 4.5
ANAMORPHIC DOOR	FOG ¼, ½, 1, 2, 3, 4, 5.					4×5	6×6	138 4.5
	DBL. FOG ⅛, ¼, ½, 1, 2, 3, 4, 5.					4×5	6×6	138 4.5
MARK II 2.35 1.85 T.V.	DIFF. 1, 2, 3, 4, 5.				T H	4×5	6×6	138 4.5
TOP LOAD/SLANT BACK	MITCH. DIFF. A, B, C, D.					4×5	6×6	138 4.5
PANAVISION H.S. MOTOR	OPTICAL FLAT					4×5	6×6	138 4.5
MITCHELL H.S. MOTOR	N.D. .30 .60 .90 STR./GRAD.					4×5	6×6	
VARIABLE SPEED	ATTENUATORS 1 STOP 2 STOP 3 STOP					4×5	6×6	
PANASPEED W/ADAP. DOOR	POLA SCREEN LRG. MT.							138 4.5
1000' MAGS.	CORAL ⅛, ¼, ½, 1, 2, 3, 4, 5.					4×5	6×6	138 4.5
400' MAGS.	STAR FILTERS 4 PT. 6 PT. 8 PT.							138 4.5
	DIOPTERS ¼, ½, 1, 2, 3. FULL/SP. LG. MT.							138 4.5
PSR/SUPER PSR 2.35 1.85 T.V.								
PANASPEED MOTOR W/COVER								
1000' MAGS. W/COVER								

PANAVISION Camera Equipment Dept.'s Order Form

Film making is truly an international activity.

Shipping Equipment Overseas

Whenever equipment is sent overseas, shipping lists detailing the equipment's description, serial number, country of origin, size, weight and value, etc., must be prepared in advance for the benefit of all the Customs Officers and Shipping Agents who are likely to be encountered along the way. The permutations of what items go together on any particular job are so vast that it is impossible to have shipping lists printed in advance but every detail is listed on PANAVISION's mainframe computer so that any shipping list is only a keyboard away.

For equipment being exported temporarily to many countries the most convenient method of coping with the problems of Import/Export is to use an International Carnet de Passage, a sort of "Passport for Camera Equipment." PANAVISION is able to supply your Shipping Agent with all the necessary information to apply for such documentation.

Carnet countries
The following Countries accept the International Carnet de Passage:

Australia
Austria
Belgium & Luxemburg
Bulgaria
Canada
Cyprus
Czechoslovakia
Denmark
Finland
France
W. Germany
Gibraltar
Greece
Hong Kong
Hungary
Iceland
Iran
Ireland
Israel
Italy

Ivory Coast
Japan
Korea
Mauritius
Netherlands
Norway
Poland
Portugal
Romania
Senegal
Singapore
South Africa
Spain
Sri Lanka
Sweden
Switzerland
Turkey
United Kingdom
United States of America
Yugoslavia

238

PANAVISION

TELEPHONE
818-881-1702
CABLE PANAVISION
TELEX 651407
PANVISION TZNA

18618 OXNARD STREET
TARZANA, CALIFORNIA USA 91356-1492

EXPORT NO: 11922
DELTA #1744 12:20pm ON 10/12/88
MAWB: 006-LAX-3466-4520-MEX
HAWB: 425735
ETA: 4:55pm ON 10/12/88

DATE: 10/19/88
PAGE: 1 OF 3

        S H I P   D I R E C T

THE FOLLOWING IS A LIST OF EQUIPMENT BEING SHIPPED TO DYNAMIC FOR
FORWARDING ON DYNAMIC AIR FREIGHT TO TRATAFILMS,S.A. C/O JOAQUIN PEREZ
CASASOLA AT AEROPUERTO INTER.DE LA DE MEXICO. THIS EQUIPMENT IS THE
PROPERTY OF PANAVISION, INC IN THE UNITED STATES TO BE USED TEMPORARILY
AND THEN RETURNED TO PANAVISION. SHIP VIA AIR FREIGHT. ALL CHARGES
COLLECT.

SHIPMENT# 1

CASE# 1   26LBS   21x14x7

                                                  CUSTOMS
    QTY SERIAL#/PRODUCT TYPE DESCRIPTION   AMOUNT
    ----
    1) STM10-560   1000' PANASTAR MAGA   1,200.00

CASE# 2   26LBS   21x14x7

                                                  CUSTOMS
    QTY SERIAL#/PRODUCT TYPE DESCRIPTION   AMOUNT
    ----
    1) STM10-213   1000' PANASTAR MAGA   1,200.00

CASE# 3   26LBS   21x14x7

                                                  CUSTOMS
    QTY SERIAL#/PRODUCT TYPE DESCRIPTION   AMOUNT
    ----
    1) STM10-150   1000' PANASTAR MAGA   1,200.00

CASE# 4   52LBS   25x22x11

                                                  CUSTOMS
    QTY SERIAL#/PRODUCT TYPE DESCRIPTION   AMOUNT
    ----
    3) IRISROD   PAIR OF IRIS RODS(P   13.50
    1) FFX-330   FOLLOW FOCUS EXTENS   27.50
    1) FXHL-229   OFFSET HANDLE   78.00

1. Portion of a PANAVISION shipping list, 2. Cover of an International Carnet de
Passage.

Camera Equipment Reminder List

As an aide-mémoire to ordering PANAVISION CAMERA EQUIPMENT the following is a general listing of the principal items of equipment that are available from PANAVISION Inc. in Los Angeles:

Cameras:

35mm CAMERAS:
PLATINUM PANAFLEX
GII GOLDEN PANAFLEX
GOLDEN PANAFLEX
PANAFLEX-X
LIGHTWEIGHT PANAFLEX for Steadicam system
PLATINUM PANASTAR
PANASTAR
SUPER PSR-200
PSR-200
PAN-ARRI III with PANAVISION lens mount

PAN ARRI IIC with PANAVISION hardfront
LOW PROFILE PAN-ARRI CAMERA
PAN-MITCHELL Mk.II with PANAVISION hardfront

65mm CAMERAS:
PANAVISION 65mm cameras for principal photography, for backgraound plates and for special effects.

16mm CAMERAS:
PANAFLEX 16

PANAFLEX camera accessories:

AATONCODE TIME CODE SYSTEM
AUTOMOBILE BASE
BALANCE PLATE
BARNIES, rain and dust covers
-lens heater type, long and short
-lens weather protectors,long and short
-magazine covers, 250, 500 and 1000 ft.
-magazine heater, 250, 500 and 1000 ft.
-weather-proof camera covers
BATTERIES:
-additional 24v Nicad type
-24v belt type
-24v solid Lead Acid type, complete with chargers
-36v
BATTERY CHARGERS
BATTERY ELIMINATOR
CAMERA REMOTE CONTROL UNIT
CHANGING BAG
CINEMATOGRAPHERS' COMPUTER PROGRAM
DIOPTERS, Full or split
-sliding type
DIRECTORS VIEWFINDERS:
-adjustable type
-PANAFINDER type
-Mitchell type

ELECTRONIC ACCESSORIES:
-field to frame synchronizer
-flash synchronizer
-line to film (50/60 Hz) synchronizer
-projector to camera synchronizer
-precision speed control (4-34 fps)
-remote on-off switch.
-remote speed control
-Stop Motion Package (PANASTAR only)
-time lapse (1/4 sec. -10 min. per frame)
-video to camera synchronizer
EXTENSION VIEWFINDER:
-intermediate length
EYEPIECE POUCH
FILTERS:
-various types and sizes
-Net frames
-Optical flats
-Sliding diffusers
-Special holders for wide-angle lenses
FOLLOW FOCUS CONTROL:
-regular type
-super type
-extension cable for above
GELATIN FILTER PUNCH
GROUND GLASS:
-additional PANAGLOW type
-matte cutter

240

HAND GRIPS and HANDLES:
-adjustable, left hand
-adjustable, right hand
-"T" type handle
-Auxiliary carrying handle to use with
 PANACLEAR
HEADS:
-PANAHEAD
-Super PANAHEAD
 -PANABALL leveler
-Worrall
 -Worrall head spacer
-O'Connor 200
-O'Connor 100
-O'Connor 50
 -O'Connor slide
 -O'Connor Ball Leveller
-Ronford 15/S
-Sachtler 7 + 7
-Sachtler 25
 -Eyepiece leveler for use with fluid heads
LENS ACCESSORIES:
-Bellows lens attachment, shift type
-Bellows lens attachment
-Distortion attachment
-Dyna Lens anti-vibration attachment
-Electronic zoom controller (additional)
-Focus control, Type FF
-Focus control high speed
-Flange focal depth gage set
-Inclining Prism
-Matte Box lens hood front flap
-Range extender, 1.5x focal length
 -2x
 -2x for Canon lenses
-Revolving mesmerizer lens
-Rifle sight attachment
-Split diopter system
-Zoom control battery complement
-Zoom lens control, foot pedal type
LOUMA CAMERA CRANE SYSTEM
MAGAZINES:
-250, 500 and 1000'
-1000' reversible (for PLATINUM PANAFLEX
 only)
-1000' high speed (for PANASTAR only)
-1000' high speed reversible (for PANASTAR
 only)
MATTE BOXES:
-standard 5.650 x 4" type
-wide-angle 5.650 x 4" type
-hand-held type
-clamp-on type
-tiltable type
-4 x 4" type (for PANAFLEX 16)
-6.6 x 6.6", 2 x grad type
-6.6 x 6.6", 3 x grad type
-6.6 x 6.6", Modular type
 -Tilting filter module
 -Dual motorized sliding grads module

-Double rotating filter module, additional
-Single rotating filter module, additional
-6.6 x 8" tiltable type
-Super sunshade extension
-Sunshade extension
PANAFADE in-shot exposure control system
PANAFLASHER in-camera negative flashing device
PANAGLIDE floating camera system
-PANAFLEX type
-Pan-Arri type
PANALENS LIGHT
PANALITE camera mounted controllable Obie light
PANALUX nightvision viewfinder
PANAROCK near ground level pan and tilt device
PANATAPE electronic tape measure
PANATATE nodal turn-over mount
PANATILT/BALANCE PLATE
PROJECTOR LENSES:
-35mm spherical
-35mm Anamorphic attachment
-70mm
RAIN and SPRAY DEFLECTORS:
-Compact type
REMOTE APERTURE CONTROL
REMOTE FOCUS AND APERTURE CONTROL UNIT
REMOTE SHUTTER LOCK CABLE
REMOTE SYNC CONTROL CABLE
RISER PLATE, 45° type
SLATE (CLAP BOARD):
-many types and sizes
SLIDING BASE PLATE
SUPER 35 modification
"T" STOP CONTROLS:
-for 5:1 & 10:1 zoom lenses
THREE PERF MOVEMENT
TILT PLATE, geared type
TRIPODS, PANAPOD:
-Standard
-Baby
 -Hi-hat
 -Spreader
UNDERWATER HOUSINGS:
-Pan-Arri type
-Anamorphic type
VIDEO ASSIST CAMERAS:
-PANAFRAME
-CCD FLICKER-FREE PANAVID
-PANAVID
 -as above, with control unit (GII & X cameras
 only)
-SUPER PANAVID 1000
-CEI type (Pan-Arri III only)
VIDEO ASSIST SYSTEMS:
-recorders
-monitors
-enhancer
-frame line generator
WATER BOX
WEATHER-PROOF COVER
WEDGE PLATE

Spherical Lens Reminder List

The decision to go Anamorphic or Spherical is one that the Producer will undoubtedly take, the choice of individual lenses is one that the Director and the Director of Photography will want to have a major say in but, no doubt, in the end it will be the Production Manager who will sign the order.

PANAVISION spherical (non-anamorphic) lenses:

PANAVISION PRIMO LENSES

Focal length	Aperture	Min. focus distance
10mm	T1.9	2'
14.5mm	T1.9	2'
17.5mm	T1.9	2'
21mm	T1.9	2'
27mm	T1.9	2'
35mm	T1.9	2'
40mm	T1.9	2'
50mm	T1.9	2'
75mm	T1.9	3'
100mm	T1.9	3'
150mm	T1.9	5'
200mm	T1.9	6'

PANAVISION PRIMO ZOOM LENS

Focal range	Aperture	Min. focus distance
17.5-75mm	T2.3	3'

PANAVISION Mk II ULTRA SPEED LENSES

Focal length	Aperture	Min. focus distance
14mm	T1.9	2'
17mm	T1.9	2'
20mm	T1.9	2'6"
24mm	T1.2	2'
29mm	T1.2	2'3"
35mm	T1.3	2'
40mm	T1.3	2'
50mm	T1.0	2'
75mm	T1.6	2'
100mm	T1.6	4'
125mm	T1.6	3'6"
150mm	T1.5	5'

PANAVISION Mk II SUPER SPEED LENSES

Focal length	Aperture	Min. focus distance
24mm	T2.0	2'
28mm	T2.0	2'
35mm	T1.6	2'
50mm	T1.4	2'6"
55mm	T1.1	2'6"

PANAVISION NORMAL SPEED LENSES

Focal length	Aperture	Min. focus distance
8mm[1]	T2.8	1'
9.8mm[2]	T2.8	2'
16mm	T2.8	1'9"
20mm	T3	2'6"
20mm	T4	2'3"
24mm	T2.8	2'3"
28mm	T2.8	2'
32mm	T2.8	2'
35mm	T2	2'
40mm	T2	2'
50mm	T2	2'6"
75mm	T2	2'9"
100mm	T2.4	3'6"
150mm	T2.8	5'

[1]Nikkor Fisheye, [2]For Pan-Arri only

CLOSE FOCUS LENSES

Focal length	Aperture	Min. focus distance
17mm	T1.9	10"
20mm	T4	8"
24mm	T2.8	8"
28mm	T2.8	8"
35mm	T2	8"

MACRO LENSES

Focal length	Aperture	Max. Magnification Ratio
40mm	T2.8	2:1
60mm	T2.8	2:1
90mm	T2.8	1:1
100mm	T2.8	1:1

CANON BNC MOUNTED LENSES

Focal length	Aperture	Min. Focus Distance	Lens type
14mm	T2.8	1'6"	
15mm	T2.8	0'7"	Fisheye
18mm[4]	T1.5	1'	
24mm[4]	T1.5	1'	
35mm[4]	T1.3	1'	
50mm[4]	T1.3	2'	
85mm[4]	T1.4	3'	
135mm	T2	4'6"	
200mm	T2.8	5'	

Suitable only for cameras with BNC lens mount.

ULTRA SPEED Mk "Z" SERIES LENSES

Focal length	Aperture	Min. Focus Distance
14mm	T1.9	2'
17mm[3]	T1.9	2'
20mm[3]	T1.9	2'6"
24mm[4]	T1.3	2'
29mm[4]	T1.3	2'
35mm[4]	T1.4	2'
50mm[4]	T1.4	2'
85mm[4]	T1.4	2'
100mm	T2	3'
135mm	T2.8	5'
180mm	T2.8	5'

SUPER SPEED Mk II "Z" SERIES LENSES

Focal length	Aperture	Min. Focus Distance
14mm	T1.9	2'
17mm[3]	T1.9	2'
20mm[3]	T1.9	2'6"
24mm[4]	T1.9	2'
29mm[4]	T1.9	2'
35mm[4]	T1.9	2'
50mm[4]	T1.9	2'
85mm[4]	T1.9	2'
100mm	T2	3'
135mm	T2.8	5'
180mm	T2.8	5'

TELEPHOTO LENSES

Focal length	Aperture	Min. Focus Distance	Lens type
200mm	T2	6'	Ultra Speed
300mm	T2.8	15'	Ultra Speed
300mm	T2.8	10'	Canon
300mm	T2	13'	Nikkor
400mm	T2.8	15'	Canon
400mm	T4	15'	
500mm	T4	23'	
600mm	T4.5	27'	Canon
800mm	T5	45'	Canon
1000mm	T6	22'	
1000mm	T9	22'	
150-600	T6	10'	Canon Vari-focal

[3]Non-Zeiss glass but color & MTF matched to complement Z series
[4]Available as a set of five lenses only.

35mm ZOOM LENSES

Focal Range	Aperture	Zoom Ratio	Optical Details	Min. Focus Distance
17.5-75mm	T2.5	4.3:1	PANAVISION PRIMO	3'
20-60mm	T3	3:1	Cooke/PANAVISION	2'3"
20-100mm	T3.1	5:1	Cooke/PANAVISION	2'6"
20-120mm	T3	6:1	Angenieux/PANAVISION	3'6"
20-125mm	T1.9	6.25:1	PANAVISION ULTRA ZOOM	4'
35-140mm	T4.5	4:1		4'
25-250mm	T4	10:1	Cooke/PANAVISION SUPER ZOOM	5'6"
25-250mm	T4	10:1	Angenieux/PANAVISION	5'6"
25-250mm	T4	10:1	Helicopter mount type	5'6"
23-460mm	T10	20:1		5'6"

Since their introduction PANAVISION anamorphic lenses have always been the most widely used.

Anamorphic Lens Reminder List

As with spherical lenses, Panavision has the widest possible range of anamorphic lenses to choose from.

PANAVISION anamorphic lenses

PRIMO SERIES COLOR-MATCHED ANAMORPHIC LENSES

Focal length	Aperture	Min. Focus Distance
35mm	T2	3'6"
40mm	T2	3'6"
50mm	T2	3'6"
75mm	T2	3'6"
100mm	T2	3'6"

"E" SERIES COLOR-MATCHED AUTO PANATAR LENSES

Focal length	Aperture	Min. Focus Distance
28mm	T2.3	5'
35mm	T2	5'
40mm	T2	5'
50mm	T2	5'
85mm	T2	5'
100mm	T2.3	5'

GOLDEN PANATAR LENSES

Focal length	Aperture	Min. Focus Distance
28mm	T2.3	3'6"
35mm	T1.4	4'6"
40mm	T1.4	5'
50mm	T1.1	4'6"

"C" SERIES LENSES

Focal length	Aperture	Min. Focus Distance
30mm	T3	4'
35mm	T2.3	2'9"
40mm	T2.8	2'6"
50mm	T2.3	2'6"
75mm	T2.5	4'6"
100mm	T2.8	4'6"
150mm	T3.5	5'
180mm	T2.8	7'

SUPER SPEED LENSES

Focal length	Aperture	Min. Focus Distance
24mm	T1.6	6'
35mm	T1.4	4'6"
50mm	T1.1	4'
50mm	T1.4	4'
55mm	T1.4	4'
75mm	T1.8	4'6"
100mm	T1.8	5'

TELEPHOTO LENSES

Focal length	Aperture	Min. Focus Distance
360mm	T4	5'6"
600mm	T4	27'
800mm	T5.6	15'
1000mm	T5.6	22'
2000mm	T9	30'

PRIMO ANAMORPHIC ZOOM LENS

Focal Range	Aperture	Zoom Ratio	Optical Details	Min. Focus Distance
35-150mm	T4	4.28:1		3'

ANAMORPHIC ZOOM LENSES

Focal Range	Aperture	Zoom Ratio	Optical Details	Min. Focus Distance
45-95mm	T4	2:1	PANAFOCAL	4'
50-95mm	T4	2:1	PANAFOCAL	4'6"
40-120mm	T4.8	3:1	Cooke/PANAVISION	2'3"
40-200mm	T4.5	5:1	Cooke/PANAVISION Super PANAZOOM	2'6"
50-500mm	T5.6	10:1	Cooke/PANAVISION Super PANAZOOM	2'6"
50-500mm	T5.6	10:1	as above, with silent zoom motor	5'6"
50-500mm	T5.6	10:1	as above, with helicopter mount motor	5'6"

SPECIAL PURPOSE ANAMORPHIC LENSES

Focal Length	Aperture	Optical Details	Min. Focus Distance
25mm	T2.5	Wide-Angle distortion lens	5'
55mm	T2.5	Macro	10"
100mm	T2.8	Insert/Process	4'6"

You're never alone with PANAVISION.

PANAVISION Representatives

Wherever in the world a film has to be made be sure there is a PANAVISION representative nearby to supply you with all your camera equipment needs.

Because the local PANAVISION representative is usually the major equipment supplier in his local area it follows that he or she can also supply you with lighting and many other of the miriads of technical items you will need to make your production go smoothly.

PANAVISION corporate headquarters:

PANAVISION INC.
18618 OXNARD ST.
TARZANA,
LOS ANGELES, CA. 91356
Tel: 818 881 1702
Fax: 818 342 8166

PANAVISION subsidiaries:

CALIFORNIA:

PANAVISION HOLLYWOOD
6779 HAWTHORN
HOLLYWOOD, CA. 90028
Tel: 213 466 3800
Fax: 213 467 0522

FLORIDA:

PANAVISION ORLANDO
2000 UNIVERSAL STUDIO PLAZA
ORLANDO, FL. 32819-7606
Tel: 407 363 0990
Fax: 407 363 0180

EUROPE:

PANAVISION EUROPE
WYCOMBE ROAD
WEMBLEY, MIDDLESEX
HAO 1QN, U.K.
Tel: (44) 1 902 3473
Fax: (44) 1 900 1557

PANAVISION representatives, U.S.A.:

CALIFORNIA:

CINE RENT WEST, INC.
991 TENNESSEE STREET
SAN FRANCISCO, CA. 94102
Tel: 415 864 4644
Fax: 415 552 9472

FLORIDA:

CINE VIDEO TECH, INC.
7330 N.E. 4TH COURT
MIAMI, FL. 33138
Tel: 305 754 2611
Fax: 305 759 2463

GEORGIA:

VICTOR DUNCAN - ATLANTA
355 ATLANTA TECHNOLOGY CENTER
1575 NORTHSIDE DRIVE NW
ATLANTA, GA. 30318
Tel: 404 355 3001
Fax: 404 355 5720

ILLINOIS:

VICTOR DUNCAN - CHICAGO
661 LA SALLE
CHICAGO, IL 60610-3770
Tel: 312 943 7300
Fax: 312 943 9043

246

MICHIGAN:

VICTOR DUNCAN - DETROIT
23801 INDUSTRIAL PARK DR., SUITE 100
FARMINGTON HILLS, MI. 48024-1132
Tel: 313 471 1600
Fax: 313 471 1940

NEW YORK:

GENERAL CAMERA CORP
540 WEST 36TH STREET
NEW YORK CITY, N.Y. 10018
Tel: 212 594 8700
Fax: 212 564 4918

TEXAS:

VICTOR DUNCAN - DALLAS
6305 N. O'CONNOR, SUITE 100
IRVING, TX. 75039-3510
Tel: 214 869 0200
Fax: 214 556 1862

PANAVISION representatives worldwide:

AUSTRALIA:

SAMUELSON FILM SERVICE (AUST.) PTY.
LTD.
1 MCLACHLAN AVENUE
P.O.BOX 542
ARTARMON, SYDNEY, N.S.W. 2064
Tel: (61) 2 436 1844
Fax: (61) 2 438 2585

SAMUELSON FILM SERVICE (AUST.) PTY.
LTD.
245-247 NORMANBY ROAD
SOUTH MELBOURNE, VICTORIA 3205
Tel: (61) 3 646 3044
Fax: (61) 3 646 4636

CANADA:

PANAVISION (CANADA) INC.
2170, AVE PIERRE-DUPUY
MONTREAL, QUEBEC H3C 3R4
Tel: (1) 514 866 7262
Fax: (1) 514 866 2297

PANAVISION (CANADA) INC.
793 PHARMACY AVENUE
SCARBOROUGH, ONTARIO M1L 3K3
Tel: (1) 416 752 7670
Fax: (1) 416 752 7599

PANAVISION (CANADA) INC.
3999 E.2ND AVENUE
BURNABY, B.C. V5C 3W9
Tel: (1) 604 291 7262
Fax: (1) 604 291 0422

CHINA:

SALON FILMS (H.K.) LTD.
C/O CHINA FILM EQUIPMENT CORP.
B25, XIN WAI STREET
BEIJING
Tel: (86) 1 201 3493
Fax: (86) 1 202 5833

FRANCE:

SAMUELSON ALGA CINEMA S.A.
24-26 RUE JEAN MOULIN
94300 VINCENNES, PARIS
Tel: (33) 1 4 328 5830
Fax: (33) 1 4 328 5077

HONG KONG:

SALON FILMS (H.K.) LTD.
MASON GARDEN BLDG. NO-2C-E
WING HING ST.,
CAUSEWAY BAY, HONG KONG
Tel: (85) 2 376 43123
Fax: (85) 2 376 43149

ITALY:

A.R.C.O. 2VIA KENNEDY, 78
ROME
Tel: (39) 6 749 0896
Fax: (39) 6 611 0152

JAPAN:

SANWA CINE EQUIPMENT
1-5-11 MOTOAKASAKA, MINATO-KU
TOKYO 107
Tel: (81) 3 404 0366
Fax: (81) 3 404 1919

TOHO COMPANY
TOHO STUDIOS
1-4-1, SEIJOO, SETAGAYA KU
TOKYO 100
Tel: (81) 3 749 4121
Fax: (81) 3 749 0330

MEXICO:

TRATAFILMS,S.A.
FELIPE VILLANUEVA NO.22
COL. GUADALUPE INN
MEXICO CITY 20, D.F.
Tel: (1) 905 680 3500
Fax: (1) 905 680 6253

PHILIPPINES:

SALON FILMS (PHILIPPINES) LTD.
66 ESTRELLA STREET
BEL-AIR VILLAGE, MAKATI
MANILA
Tel: (63) 2 88 16 91
Fax:

SPAIN:

CAMARA RENT INTERNATIONAL
MAURICIO LEGENDRE, 36
MADRID-16
Tel: (34) 1 733 4131
Fax: (34) 1 315 6036

CAMARA RENT INTERNATIONAL
PASAJE DE ARAGON, 3
BARCELONA-11
Tel: (34) 3 323 2491

SWEDEN:

FILM AB RUNE ERICSON
HUVUDSTAGATAN 12
S 171 58 SOLNA, STOCKHOLM
Tel: (46) 8 734 0845
Fax: (46) 8 734 9118

THAILAND:

SALON FILMS (THAILAND) LTD.
45 SOILANG SUAN, STE. 16
BANGKOK
Tel: (66) 2 252 8591
Fax: (66) 2 255 1419

U.K.:

SAMUELSON FILM SERVICE (LONDON) LTD.
21 DERBY RD., GREENFORD
LONDON, UB6 8UJ
Tel: (44) 1 578 7887
Fax: (44) 1 578 2733

WEST GERMANY:

F.G.V. SCHMIDLE & FITZ
ROTBUCHENSTRASSE, 1B
8000 MUNICH 90
Tel: (49) 89 690 7070
Fax: (40) 89 699 0466

248

PANAVISION
Credits

A proud list.

Pictures Photographed Using PANAVISION Cameras and/or Lenses

Note: The following list is of titles known to Panavision and has been compiled using the records available. It is subject to amendment and correction. In some cases the titles may have been changed after production commenced. Anyone who has been involved with a particular production and sees that it is misstated or omitted or notes any other correction or addition that might be made to this list is asked to please send a note about it to PANAVISION Inc., 18618 Oxnard St., Tarzana, CA 91358, U.S.A.

This list does not include films made solely for television presentation.

PANAVISION credits:

"10"
1776
1918
1941
1969
2010
10 E MIA SORELLA
11 HARROWHOUSE
2001 - A SPACE ODYSSEY
25th HOUR
3 PLACES POUR LE 26
32 DICEMBRE
40 CARATS
50/50
52 PICK UP
633 SQUADRON
8 BELLS TOLL
8 MILLION WAYS TO DIE
84 CHARING CROSS ROAD
99 AND 44/100 % DEAD
9½ WEEKS
ABBA - THE MOVIE
ABBESS, The
ABDICATION, The
ABDULLAH
ABOUT LAST NIGHT
ABSENCE OF MALICE
ABSOLUTION
ACCEPTABLE RISK
ACCIDENTAL TOURIST
ACE ELI AND ROGER OF THE
  SKIES
ACES GO PLACES III
ACES HIGH
ACORN PEOPLE
ACQUE DI PRIMAVERA
ACROSS 110th STREET

ACTS OF PETER AND PAUL,
  The
ADA
ADAM'S RIB
ADAM'S WOMAN
ADIOS AMIGO
ADVANCE OF RABBI JACOBS
ADVANCE TO THE REAR
ADVENTURERS, The
ADVENTURES IN
  BABYSITTING
ADVENTURES OF BUCKEROO
  BONZAI, The
ADVENTURES OF GERARD,
  The
ADVENTURES OF HUCKLE-
  BERRY FINN, The
ADVENTURES OF THE LAMP
ADVISE AND CONSENT
AFRICAN ELEPHANT
AFTER THE FOX
AGAINST ALL ODDS
AI FUTATABI
AI KONNICHIWA
AI NO KANE
AI TABIDACHI
AIDA
AIFUTATABI
AIJO MONOGATARI
AIR AMERICA
AIR FORCE
AIRPLANE
AIRPLANE II
AIRPORT
AIRPORT 1975
AIRPORT 1977
AIWA CROSSOVER

AKURYOU TOU
AL MAGEN DE LA VIDA
AL MARGA DE LA VIDA
ALAMO
ALASKA MONOGATARI
ALDO ET JUNIOR
ALEX IN WONDERLAND
ALFIE DARLING
ALFRED THE GREAT
ALICE DOESN'T LIVE HERE
  ANYMORE
ALIEN
ALIVE, ALIVE O
ALL AMERICAN BOY
ALL AMERICAN GIRL
ALL CREATURES GREAT &
  SMALL
ALL FALL DOWN
ALL IS BUT TOYS
ALL NIGHTER
ALL THAT JAZZ
ALL THE FINE YOUNG
  CANNIBALS
ALL THE PRESIDENT'S MEN
ALL THINGS BRIGHT AND
  BEAUTIFUL
ALPHABET CITY
ALTERED STATES
ALUCINEMA
ALVEREZ KELLY
AMADEUS
AMER BRUJO, El
AMERICAN ANTHEM
AMERICAN DATE
AMERICAN DREAMER
AMERICAN FLYER
AMERICAN GRAFFITI

250

251

252

CULPEPPER CATTLE CO., The
CURÉE, La
CURSE OF DARK SHADOWS
CURSE OF THE PINK
  PANTHER, The
CUTTING CLASS
CYBORG

D.A.R.Y.L.
DAD
DADDY'S GONE A HUNTING
DADDY'S LITTLE GIRL
DAKOTA
DALLAS
DALTON
DAMIEN - OMEN II
DAMN THE DEFIANT
DAMNATION ALLEY
DANCE
DANCE TWO
DANCES WITH WOLVES
DANDY IN ASPIC, A
DANGEROUS CHARTER
DANGEROUS DAVIES
DANGEROUS GAME
DANIEL
DANNY, CHAMPION OF THE
  WORLD
DANY, THE ALL-AMERICAN
  GIRL
DARK, The
DARK AGE
DARK ANGEL
DARK BEFORE DAWN
DARK CRYSTAL
DARK MISSION
DARK OF THE SUN
DARLING
DARLING LILI
DAS HAUS IN MONTEVIDEO
DATE NIGHT
DAY BEFORE, The
DAY FOR NIGHT
DAY MARS INVADED THE
  EARTH, The
DAY OF A CHAMPION
DAY OF THE DOLPHIN
DAY OF THE EVIL GUN
DAY OF THE JACKAL, The
DAY OF THE LOCUST
DAY OF THE TRIFFIDS, The
DAY THE CLOWN CRIED, The
DAY TIME ENDED, The
DAYS OF HEAVEN
DEAD AS THEY COME
DEAD BANG
DEAD CALM
DEAD CERT
DEAD KIDS
DEAD POETS' SOCIETY
DEAD POOL
DEAD RINGERS
DEAD SOLID PERFECT
DEADHEAD MILES
DEADLINE

DEADLY BLESSINGS
DEADLY COMPANIONS
DEADLY FORCE
DEADLY HONEYMOON
DEALING
DEATH CAME SUDDENLY
DEATH FLASH
DEATH HUNT
DEATH IN VENICE
DEATH JOURNEY
DEATH OF A SOLDIER
DEATH ON THE NILE
DEATH TRAP
DEATH WATCH
DEATH WISH
DECEPTION
DECEPTIONS
DECOY, The
DEEP, The
DEEP IS THE FOREST
DEER HUNTER, The
DEFENSE PLAY
DEJA VU
DELICATE BALANCE, A
DELIVERANCE
DELUGE
DEMON SEED
DEMON WARRIOR
DEPART, Le
DEPTH OF FEELING
DERBORENCE
DESCENTE AUX ENFERS
DESERT BLOOM
DESERT JOURNEY
DESERTER, The
DETECTIVE, The
DETROIT 9000
DEUXIEME RECONTRE, Le
DEVIL'S ANGELS
DEVIL'S BACKBONE, The
DEVIL'S BRIGADE, The
DEVIL'S EXPRESS, The
DEVIL'S WIDOW
DEVILED HAM
DEVILS, The
DIAMOND CUT DIAMOND
DIAMOND HEAD
DIAMONDS ARE FOREVER
DICK & JANE
DIE BOSS, DIE QUICKLY
DIE HARD
DIMBOOLA
DION BROTHERS, The
DIRTY DANCING
DIRTY DINGUS MAGEE
DIRTY HARRY
DIRTY LITTLE BILLY
DISTANT THUNDER
DISTANT TRUMPET, A
DIVINAS PALABRAS
DIVINE MADNESS
DOC HOOKER'S BUNCH
DOC SAVAGE
DOCKPOJKEN

DOCTOR, YOU'VE GOT TO BE
  KIDDING
DOCTOR ZHIVAGO
DOG DAY AFTERNOON
DOG SOLDIERS
DOLL'S HOUSE, A
DOMANI, Da
DOMINICK & EUGENE
DOMINO PRINCIPLE
DON'S PARTY
DON'T ASK ME
DON'T LOOK NOW... WE'RE
  BEING SHOT AT
DON'T MAKE WAVES
DON CAMILLO
DON GIOVANNI
DON QUIXOTE
DONG
DOOUBLE EXPOSURE
DORADO, El
DOUBLE DARE &
  NICKELODEON
DOUBLE DEAL
DOUBLE RANSOM
DOUBLE TROUBLE
DOVE, The
DOWN AND OUT IN BEVERLY
  HILLS
DOWNTOWN
DOWNTOWN HERO
DR. GEORGE PRATORIUS
DR. GOLDFOOT AND THE BI-
  KINI MACHINE
DRABBLE
DRACULA (1979)
DRAGON FLIES, The
DRAGONSLAYER
DRAMMA DA CAMERA
DREAM DEMONS
DREAM TEAM
DREAM WISH
DREAMS
DRESS, The
DRESSED TO KILL
DRIVE-IN
DRIVER, The
DROP DEAD DARLING
DROWNING POOL, The
DRUGSTORE COWBOY
DRUM
DUBIOUS PATRIOTS
DUCHESS AND THE DIRTWA-
  TER FOX, The
DUEL
DUEL AT EZO
DUELLISTS, The

E.T. THE EXTRA-TERRESTRIAL
EAGLE'S WING
EAGLE HAS LANDED, The
EARTH GIRLS ARE EASY
EARTHLING, The
EARTHQUAKE
EAT A BOWL OF TEA
EBENEZA SCROOGE

254

ECCO
EFFECT OF GAMMA RAYS ON
MAN IN THE MOON
MARIGOLDS
EGYPTIAN PROJECT
EIENNO 1/2
EIGA JOYUU
EIGER SANCTION, The
EIGHT IS ENOUGH
EIGHTEEN IN THE SUN
ELAN PROJECT FOUR
ELECTRA GLIDE IN BLUE
ELECTRIC HORSEMAN
ELEPHANT MAN, The
ELLIS ISLAND
ELVIRA - MISTRESS OF THE
DARK
ELVIS- THAT'S THE WAY IT IS
ELYSIAN FIELDS
EMBRYO
EMERALD
EMERALD FOREST, The
EMERALD ISLE, The
EMILIANO ZAPATA
EMMANUEL - JOYS OF A
WOMAN
EMPEROR AND THE GENERAL,
The
EMPEROR OF THE BRONX
EMPIRE OF THE SUN
EMPIRE STRIKES BACK, The
END, The
ENEMY MINE
ENEMY OF THE PEOPLE, An
ENFANCE DE L'ARI, L'
ENFORCER, The
ENGLAND MADE ME
ENSIGN PULVER
ENTEBBE: OPERATION
THUNDERBOLT
ENTER LAUGHING
ENTER THE DRAGON
ENTITY, The
EQUILIBRI
EQUIS
ERIK THE VIKING
ERNEST SAVES CHRISTMAS
ESCAPE
ESCAPE FROM ALCATRAZ
ESCAPE FROM ATHENA
ESCAPE FROM NEW YORK
ESCAPE FROM THE PLANET
OF THE APES
ESCAPE FROM ZAHRAIN
ESCAPE TO VICTORY
ESCARGOT NOIR, L'
ESQUILACHE
ESTOUFFODE AUX CARIABES
ETAT D'AME
ETERNITY
ETINCELLE, L'
EVERBODY'S ALL-AMERICAN
EVERY LITTLE CROOK &
NANNY
EVERYBODY WINS

EVICTORS, The
EVIL ANGELS
EVIL GUN
EVIL UNDER THE SUN
EXECUTIONER, The
EXODUS
EXORCIST
EXPLORERS
EXTRAORDINARY SEAMAN,
The
EYE FOR AN EYE, An
EYES OF LAURA MARS, The
EYEWITNESS

F.I.S.T.
F.J. HOLDEN, The
F/X
FABEINNE
FACE THE WIND
FALL OF THE ROMAN EMPIRE,
The
FALLING IN LOVE
FALSE FACE
FAME
FAMILY , The
FAMILY BUSINESS
FAN, The
FANNY
FANTASTIC VOYAGE
FAR FROM THE MADDING
CROWD
FAR NORTH
FAR WEST, La
FAREWELL MY LOVELY
FAREWELL TO THE KING
FARMERS, The
FASCINATION, La
FAST TIMES AT RIDGMONT
HIGH
FAT CITY
FATAL ATTRACTION
FATAL BEAUTY
FATAL CHARM
FATAL VISION
FATE
FATHER CLEMENTS
FATHER DOWLING
FAUSSES CONFIDENCES, Les
FBI STORY, The
FEAR IS THE KEY
FEARLESS VAMPIRE KILLERS
FEDORA
FELLINI SATYRICON
FEMALE ANIMAL
FEMME DE MON POTE, La
FERRIS BUELLERS DAY OFF
FIDDLER ON THE ROOF
FIGURES IN A LANDSCAPE
FINAL CHAPTER, The
FINAL CONFLICT, The
FINAL COUNTDOWN, The
FINALE IN BERLIN
FINE GOLD
FINE MESS, A
FINE PAIR, A

FINE ROMANCE, A
FINIAN'S FAINBOW
FINITE BLONDE, Le
FIRE AND ICE
FIRE POWER
FIRE SALE
FIREBALL 500
FIRECREEK
FIREFOX
FIREPOWER
FIRST BORN
FIRST DATE
FIRST LOVE
FIRST MEN IN THE MOON
FIRST MONDAY IN OCTOBER
FIRST NUDIE MUSICAL
FIRST TO FIGHT
FISH CALLED WANDA, A
FISHER KING
FISTS OF STEEL
FITZWILLY
FIVE EASY PIECES
FLAGRANT DESIR
FLAME
FLAMENCO
FLAMENCO BLUES
FLAMINGO KID
FLAP
FLASHBACK
FLASHDANCE
FLEA IN HER EAR, A
FLETCH
FLETCH II
FLEURS DU MIEL, LES
FLIGHT FROM ASHIYA
FLIM FLAM MAN
FLOWER DRUM SONG
FLOWERS IN THE CRANNIED
WALL
FLY, The
FLY II, The
FLYING
FODRINGSÄGARE
FOG, The
FOLLOW ME
FOLLOW THAT BIRD
FOLLOW THAT DREAM
FOLLOW THE BOYS
FOLLOWING THE FOOTSTEPS
FOOTLOOSE
FOR KEEPS
FOR PETE'S SAKE
FOR QUEEN AND COUNTRY
FOR YOUR EYES ONLY
FORBIN PROJECT, The
FORCE 10 FROM NAVARONE
FORD FAIRLANE
FOREIGN BODY
FORFEIT
FORJA DE UN REBELDE
FORMULA, The
FORTUNE, The
FORTUNE COOKIE, The
FORTY POUNDS OF
TROUBLE

GUNS OF NAVARONE
GUNS OF THE MAGNIFICENT
  SEVEN
GUTS & GLORY
GWENDOLINE

HABIT
HAIL
HAIR
HAIR PIN CIRCUS
HAKKOUDASAN
HAKO KIRAMEKUHATE
HALF A SIXPENCE
HALL OF MIRRORS
HALLELUJAH TRAIL
HALLOWEEN
HALLOWEEN II
HALLOWEEN III
HALLOWEEN IV
HAMBONE & HILLIE
HAMBURGER - THE MOTION
  PICTURE
HAMBURGER HILL
HANAMO ARASHIMO
  TORAJIRO
HANANO RAN
HANAUMA BAY
HANGED MAN, The
HANNAH
HANNAH AND HER SISTERS
HANNIE CAULDER
HANOVER STREET
HAPPY BIRTHDAY, GEMINI
HAPPY ENDING, The
HAPPY HOOKER
HAPPY NEW YEAR
HAPPY TOGETHER
HARBOUR LIGHTS
HARD CONTRACT
HARD RAIN
HARD TIMES
HARE TOKIDOKI SATSUJIN
HAREM
HARLEQUIN
HARLOW
HAROLD & MAUDE
HARPER
HARPER VALLEY P.T.A.
HARRY & THE HENDERSONS
HARRY & TONTO
HARRY AND WALTER GO TO
  NEW YORK
HARRY IN YOUR POCKET
HARUKURU ONI
HARUM SCARUM
HAUNTED, The
HAUNTED HONEYMOON
HAUNTED PALACE, The
HAUNTING, The
HAUNTING OF JULIA, The
HAWAII
HAWAIIAN DREAM
HAWAIIANS, The
HAWKIN'
HAWMPS

HAZARD OF HEARTS, A
HE'S NOT YOUR SON
HEAD OF THE DRAGON
HEAD OFFICE
HEALTH
HEART OF DIXIE
HEARTBEEPS
HEARTBREAK KID, The
HEARTBREAK RIDGE
HEARTBURN
HEARTS ARE TRUMPS
HEARTS OF THE WEST
HEAVEN'S GATE
HEAVEN CAN WAIT
HEAVEN WITH A GUN
HELL'S ANGELS
HELL BENT FOR LEATHER
HELL IN THE PACIFIC
HELLBOUND
HELLFIGHTERS
HELLO AGAIN
HELLRAISER
HENRY V (1988)
HENRY V (1989)
HER ALIBI
HERETIC, The
HERITIER, L'
HERO, The
HERO'S ISLAND
HERO AIN'T NOTHING BUT A
  SANDWICH, A
HERO WORK
HEROES
HEROIC PIONEERS
HEROINE, L'
HEROS OF TELEMARK
HHANAI MORI
HIDE IN PLAIN SIGHT
HIDING PLACE, The
HIGH ANXIETY
HIGH COUNTRY, The
HIGH MOUNTAIN RANGER
HIGH PLAINS DRIFTER
HIGH SCHOOL U.S.A.
HIGH TIDE
HIGH VELOCITY
HIGHLANDER
HIGHTEEN BOOGIE
HIMATSURI
HINDENBERG, The
HIRELING, The
HIS MAJESTY'S SARGENT
HISSATSU, The
HISSATSU, The - URAKA
  OMOTEKA
HISSATSU - BURANKANNO
  KAIBUTSUTACHI
HISSATSU 4
HISTORY OF THE WORLD
  PART I
HIT, The
HITCHER, The
HITLER'S S.S.
HOLCROFT COVENANT, The
HOLD ON!

HOLD UP
HOLE IN THE HEAD
HOMBRE
HOME FROM THE HILL
HOME OF THE BRAVE
HOME TOWN USA
HOMEBOY
HOMECOMING, The
HOMEROOM NEWS
HOMME A REDRESSER, Un
HOMME QUI N'ETAIT PAS LA,
  L'
HOMME QUI VIVAIT AU RITZ,
  L'
HONEY, I SHRUNK THE KIDS
HONEYMOON HOTEL
HONEYMOON MACHINE, The
HONEYMOON SUITE
HONEYSUCKLE ROSE
HONKY
HONOR THY FATHER
HOOK, The
HOOPER
HOPSCOTCH
HORIZONTAL LIEUTENANT
HORRIBLE DR. HITCHCOCK
HORS LA LOI
HORSEMEN, The
HOSHIZUNA MONOGATARI
HOSPITAL, The
HOT POTATO
HOT PURSUIT
HOT ROCK, The
HOT STUFF
HOT TARGET
HOT TO TROT
HOTEL
HOTEL DE LA PLAGE, L'
HOTEL PARADISO
HOUNDS OF THE
  BASKERVILLES (1974), The
HOUR OF THE GUN
HOUSE
HOUSE BY THE CEMETERY,
  The
HOUSE CALLS
HOUSE MADE OF DAWN
HOUSE OF GAMES
HOUSE OF HAMMER, The
HOUSE OF THE DAMMED
HOUSEKEEPING
HOW SWEET IT IS
HOW THE WEST WAS WON
HOW TO DESTROY THE MOST
  SUCCESSFUL SECRET
  AGENT IN THE WORLD
HOW TO SAVE A MARRIAGE
  AND RUIN YOUR LIFE
HOW TO STEAL A MILLION
HOW TO STUFF A WILD BIKINI
HOW TO SUCCEED IN BUSI-
  NESS WITHOUT REALLY
  TRYING
HOWARD THE DUCK
HUCKLEBERRY FINN (1974)

257

HUD
HUMAN FACTOR, The
HUNCHBACK
HUNGER, The
HUNK
HUNT FOR RED OCTOBER
HURRY SUNDOWN
HUSTLER, The
HYDE
HYORYO KYOSHITSU

I, THE JURY
I'M DANCING AS FAST AS I
  CAN
I COULD GO ON SINGING
I ESCAPED FROM DEVIL'S
  ISLAND
I LOVE NEW YORK
I LOVE YOU
I NEVER PROMISED YOU A
  ROSE GARDEN
I WALK THE LINE
I WANT TO HOLD YOUR HAND
I WILL, I WILL... FOR NOW
ICE CASTLES
ICE STATION ZEBRA
ICEMAN
ICEMAN COMETH, The
IDEA OF AMERICA
IF TOMORROW COMES
IHMISELSON IHANUUS JA
  KURJUUS
IKOKA MODOROKA
ILLEGALLY YOURS
ILLUSTRATED MAN, The
IMAGES
IMAGINE JOHN AND YOKO
IMAGO URBIS
IMMEDIATE FAMILY
IMPOSSIBLE YEARS, The
IMPULSE
IN COLD BLOOD
IN COUNTRY
IN HARM'S WAY
IN SEARCH OF ANNA
IN THE COOL OF THE DAY
IN THE DEVIL'S GARDEN
IN THE HEAT OF THE NIGHT
IN THE NAME OF THE LAW
IN THIS HOUSE OF BREDEF
INAUGURAL BALL
INCHON
INCREDIBLE SARAH, The
INDEPENDENCE
INDIANA JONES AND THE
  LAST CRUSADE
INDIANA JONES AND THE
  TEMPLE OF DOOM
INDIO
INDISCREET
INDONISIA
INFRA - MAN
INNERSPACE
INSIDE DAISY CLOVER
INSIDE MOVES

INSIDE OUT
INSIDE THE THIRD REICH
INSPECTOR CLOUSEAU
INTENT TO KILL
INTERIORS
INTERMEDIA
INVADERS FROM MARS (1986)
INVITATATION TO THE
  WEDDING
IRMA LA DOUCE
IRONWEED
IRRECONCILABLE
  DIFFERENCES
IS PARIS BURNING?
ISABELLA OF SPAIN
ISADORA
ISHTAR
ISLAND, The
ISLAND OF DR. MOREAU
ISLAND OF LOVE
ISLANDS IN THE STREAM
IT'S A MAD, MAD, MAD, MAD
  WORLD
IT'S A TEST!
IT'S ALIVE
IT HAD TO BE YOU
IT HAPPENED AT THE
  WORLD'S FAIR
IT STARTED WITH A KISS
ITALIAN JOB, The
ITOSHIKI HIBI
ITOSHINO ELLY
ITOSHINO HALF MOON
IZAKAYA CHOJI

J.W. COOP
JACK OF DIAMONDS
JACK THE RIPPER
JACKALS
JACOB'S LADDER
JAD
JAGGED EDGE
JAILHOUSE ROCK
JANUARY MAN
JARDIN DE TIA ISABEL, El
JAWS
JAWS '87
JAWS II
JAZZ BOAT
JAZZ SINGER (1980)
JEALOUSY GAME
JEDE MENGE KOHLE
JEMINI Y & S
JEREMIAH JOHNSON
JEREMIAH OF JACOBS NECK
JERICHO
JESSICA
JESUS OF NAZARETH
JIGOKUHEN
JIGSAW MAN, The
JISHIN RETTOU
JO JO DANCER
JOANNA
JOE KIDD
JOE VS. THE VOLCANO

JOGGER, The
JOHN AND MARY
JOHNNY BE GOOD
JOHNNY FIRECLOUD
JOHNNY GOT HIS GUN
JOLSON STORY
JONATHAN LIVINGSTON
  SEAGULL
JONQUE CHINOISE, La
JOSEPH ANDREWS
JOURNEY THROUGH
  ROSEBUD
JOY IN THE MORNING
JOYEUSES PAQUES
JOYRIDERS
JUDAS PROJECT
JUDITH
JUGE ET L' ASSASSIN, LE
JUGGERNAUT
JUILLET EN SEPTEMBRE
JULIA
JULIA (Denmark)
JULIET GAME
JULIUS CAESAR (1970)
JUMBO
JUMEAU, Le
JUMPING JACK FLASH
JUST A PIECE OF PAPER
JUST TELL ME WHAT YOU
  WANT
JUSTINE

K-9
KAETTE KITA WAKADAISHOU
KAGEMUSHA
KAGI
KAIKYOU
KAMIKAZE
KAMOURASKA
KANAKONO TAMENI
KANASHII IROYANEN
KANGAROO
KANSAS
KANSAS CITY BOMBER
KARATE KID
KARATE KID II
KARATE KID III, The
KASEKI NO MORI
KATAKUNO HITO
KATAYOKUDAKENO TENSHI
KAZABLAN
KAZENO UTAO KIKE
KEATON'S COP
KEEP, The
KELLY'S HEROES
KEUFS, Les
KEY EXCHANGE
KEY WITNESS
KISS OF THE SPIDER WOMAN
KHARTOUM
KICKBACK, The
KICKBOXER
KID BLUE
KID BROTHER, The
KID GLOVE
KIDNAPPED

KILL HIM IN AMSTERDAM
KILLER CLOWNS FROM
   OUTER SPACE
KILLER ELITE
KILLER INSIDE ME
KILLERS OF KILMANJARO
KILLING DEVICE, The
KILLING IN THE DARK
KIMURAKENO HITOBITO
KINDRED, The
KINEMANO TENCHI
KING DAVID
KING GUN
KING KONG (1976)
KING OF OLYMPICS, The
KING OF THE GYPSIES, The
KINGEN KAKUMEI
KINGS OF THE SUN
KISS ME, STUPID
KISS THE OTHER SHEIK
KISSIN' COUSINS
KLUTE
KNIGHTS AND EMERALDS
KNOW YOUR OWN RABBIT
KOI NO BRANKO
KOIBITOTACHINO JIKOKU
KONO AINO MONOGATARI
KONO MUNENO TOKIMEKI
KONOKONO NANATSUNO
   OIWAINI
KOREGA DOSADA
KORPENS SKUGGA
KOUYA HIJIRI
KRAKATOA, EAST OF JAVA
KRAMER VS. KRAMER
KREMLIN LETTER, The
KRULL
KRYPSKYTTERE
KUCHIBUE O FUKU TORAJIRO
KURANO NAKA
KUROI AME
KUROI DRESS NO ONNA

LADY CAROLINE LAMB
LADY FROM YESTERDAY, The
LADY HAMILTON
LADY ICE
LADY IN CEMENT
LADY IN THE CAR WITH
   GLASSES AND A GUN
LADY JANE GREY
LADY L
LADY MOBSTER
LADY SINGS THE BLUES
LAGENS NAMN, I
LAIR OF THE WHITE WORM
LAKE OF DRACULA
LANCELOT AND GUINEVERE
LAND'S END
LAS VI BORAS TAMBIEN CAM-
   BIAN DE PIEL
LASSIE MY LASSIE
LAST AMERICAN HERO (1973),
   The
LAST CHALLENGE, The

LAST DAYS OF PATTON, The
LAST DETAIL, The
LAST DRAGON, The
LAST FRONTIER, The
LAST GRENADE , The
LAST HARD MAN, The
LAST IMAGE, The
LAST OF SHELIA, The
LAST PICTURE SHOW, The
LAST REMAKE OF BEAU
   GESTE
LAST RITES
LAST RUN, The
LAST STARFIGHTER, The
LAST TEN DAYS
LAST TEN DAYS OF HITLER,
   The
LAST TYCOON, The
LATE LIZ, The
LATE SHOW, The
LAW & ORDER
LAW AND DISORDER
LAW AND TOMBSTONE
LAWRENCE OF ARABIA
LE MANS
LEAN ON ME
LEARNING TREE, The
LEGAL EAGLES
LEGEND OF LYLAH CLARE,
   The
LEGEND OF THE HOLY
   DRINKER, The
LEGEND OF THE LONE
   RANGER, The
LEGEND OF THE SEVEN
   GOLDEN VAMPIRES, The
LEGENT
LEMON SISTERS
LEONARD VI
LEONSK: INCIDENT, The
LEPKE
LES PATTERSON SAVES THE
   WORLD
LET'S GET HARRY
LET'S SPEND THE NIGHT
   TOGETHER
LETHAL WEAPON
LETHAL WEAPON II
LEVI ET GOLIATH
LEVIATHAN
LICENSE REVOKED
LIEN DE PARENTE, Le
LIFE AFTER LIFE
LIFE AHEAD, A
LIFE AND TIMES OF JUDGE
   ROY BEAN, The
LIFEGUARD
LIGHT AT THE EDGE OF THE
   WORLD
LIGHT HORSEMEN
LIGHT IN THE PIAZZA
LIGHT OF DAY
LIGHTHOUSEMEN, The
LIGHTS OF INCHON
LIKE FATHER, LIKE SON

LILLIES OF THE FIELD
LIME STREET
LIMIT UP
LION IN WINTER, The
LION OF THE DESERT, The
LIONHEART
LIPSTICK
LIQUIDATOR
LISTOMANIA
LITTLE ARK, The
LITTLE BIG MAN
LITTLE DARLINGS
LITTLE FAUS AND BIG
   HALSY
LITTLE GIRL WHO LIVES
   DOWN THE LANE, The
LITTLE LORD FAUNTLEROY
LITTLE MALCOLM
LITTLE MONSTERS
LITTLE NIGHT MUSIC, A
LITTLE NIKITA
LITTLE SHOP OF HORRORS
   (1985)
LITTLE SISTER
LIVE A LITTLE, LOVE A LITTLE
LIVE A LITTLE, STEAL A LOT
LIVING DAYLIGHTS
LIVSFARLIG FILM
LOCO VENENO
LOGAN'S RUN
LOI SAUVAGE, La
LOLLY - MADONNA XXX
LOMBARD & GABLE
LONDON
LONDON AFFAIR, The
LONDON EXPERIENCE, The
LONDON HEATHROW
LONE WOLF
LONELY HEARTS, The
LONG DUEL, The
LONG ESCAPE, The
LONG GOODBYE, The
LONG LOST FRIEND, A
LONG RIDE HOME, The
LONG WALK HOME
LONG WAY FROM HOME, A
LONG WEEKEND, The
LONGEST DAY, The
LONGS MANTEAUX, Les
LONLEY ARE THE BRAVE
LOOKER
LOOKING FOR LOVE
LOOKING FOR MR. GOODBAR
LOOKING GLASS WAR
LORD HIGH EXECUTIONER
LORD JIM
LORD OF THE FLIES (1989)
LORD SHANGO
LOST BOYS
LOST COMMAND (1966)
LOST HORIZON (1973)
LOST IN AMERICA
LOST IN THE STARS
LOST MAN, The
LOUPS ENTRE EUX, Les

259

MOHAMMED ALI, THE GREATEST
MOI VOULOIR TOI
MOIS D'AVRIL....., Les
MOLLY MAGUIRES, The
MOMENT AT LAST, A
MOMENT BY MOMENT
MONEY PIT
MONEY TRAP, The
MONSIEUR HIRE
MONSIGNOR QUIXOTE
MONSTER CLUB, The
MONSTER SQUAD
MONTE CARLO
MONTE WALSH
MONTY CARLO OR BUST
MONTY PYTHON AND THE HOLY GRAIL
MOON 44
MOON IN THE GUTTER, The
MOON OVER PARADOR
MOON SHOT
MOONRAKER
MOONSHINE WAR, The
MOONSTRUCK
MOONTRAP
MORE AMERICAN GRAFFITI
MORE THINGS CHANGE, The
MORFALOUS, Les
MORGAN THE PIRATE
MORNING AFTER, The
MOROCCO 7
MORT D'UN BUCHERON, La
MORTUARY ACADEMY
MOSCOW ON THE HUDSON
MOSQUITO COAST, The
MOST DANGEROUS MAN IN THE WORLD, The
MOTHER JUGS AND SPEED
MOTHERS
MOUNT KAKKODA
MOUNTAIN KING, The
MOUNTAIN MEN
MOUSTACHU, Le
MOUTARDE ME MONTE AU NEZ, LA
MOVE
MOVE OVER DARLING
MOVING
MOVING TARGET
MR. MAJESTYK
MR. PRESIDENT
MR. QUILP
MR. RICCO
MRS. BROWN YOU'VE GOT A LOVELY DAUGHTER
MRS. MANNING'S WEEKEND
MUD
MUJERES AL BORDE DE UN ATAQUE DE NERVIOS
MUNECA REINA
MUPPETS TAKE MANHATTEN, The
MURDER BY DEATH

MURDER BY DECREE
MURDER ON THE ORIENT EXPRESS
MURDER WEAPON
MURDER WITH MIRRORS
MURPHY'S LAW
MURPHY'S ROMANCE
MURPHY'S WAR
MURROW
MUSCLE BEACH PARTY
MUSIC LOVERS, The
MUTINY ON THE BOUNTY (1962)
MY BLOOD RUNS COLD
MY BRILLIANT CAREER
MY FAIR LADY
MY FAVORITE YEAR
MY MOTHER THE GENERAL
MY NAME IS NOBODY
MY PLEASURE IS MY BUSINESS
MY SCIENCE PROJECT
MY SIDE OF THE MOUNTAIN
MY STEPMOTHER IS AN ALIEN
MY SUMMER VACATION
MYERLING, The
MYRA BREKINRIDGE
MYSTIC PIZZA

NADINE
NAIROBI
NAKED APE, The
NAKED PREY, The
NANIWANO KOINO TORAJIRO
NAPOLEON ET JOSEPHINE
NARROW MARGIN
NASHVILLE
NASTY HABITS
NATIONAL LAMPOON'S ANIMAL HOUSE
NATIONAL LAMPOON'S XMAS VACATION
NATURAL, The
NAVIGATOR, The
NAVY SEALS
NEGRO CON UN SAXO, Un
NELSON AFFAIR, The
NEMO
NEPTUNE FACTOR, The
NETWORK
NEVADA SMITH
NEVER CRY WOLF
NEVER SAY NEVER AGAIN
NEVER SO FEW
NEVER TOO LATE
NEW CENTURIONS, The
NEW LIFE, A
NEW YORK, NEW YORK
NEW YORK STORIES
NEWSFRONT
NEXT OF KIN
NEXT STOP GREENWICH VILLAGE
NICKELODEON

NICHOLAS AND ALEXANDRA
NIGHT CROSSING
NIGHT IN THE LIFE OF JIMMY REARDON, A
NIGHT MOVES
NIGHT OF LOVE, A
NIGHT OF THE GENERALS, The
NIGHT OF THE QUARTER MOON
NIGHT WATCH
NIGHTHAWKS
NIGHTMARE ON ELM STREET
NIGHTMARE ON ELM STREET PART 4
NIGHTMARE RALLY
NIGHTSIDE
NIJINSKI
NIJYUYONNO HITOMI
NINE HOURS TO RAMA
NINGEN KAKUMEI
NINJA
NINJA ACADEMY
NINTH CONFIGURATION, The
NO BEAST SO FIERCE
NO BLADE OF GRASS
NO HOLDS BARRED
NO ISHIBUMI
NO LONGER ALONE
NO MERCY
NO SURRENDER
NO WAY TO TREAT A LADY
NOBEL HOUSE
NOBODY'S FOOL
NOCTURNE INDIEN
NOGARENO MACHI
NOMUGI TOUGE
NONE BUT THE BRAVE
NORIKOWA IMA
NORMA RAE
NORMAN, IS THAT YOU
NORMAN VINCENT PEALE
NORSEMAN, The
NOSFERATU
NOT ON YOUR LIFE
NOT WITH YOU, NOT WITHOUT YOU
NOTHING IN COMMON
NOW I LAY ME DOWN TO SLEEP
NOYADE INTERDITE
NUIT AMERICAINE, La
NUIT BENGALI, La
NUNZIO
NUTS

O'CONNORS, The
O.S.S.
OBSESSION
OCEAN'S ELEVEN
OCTOPUSSY
ODD COUPLE, The
ODE TO BILLIE JOE
ODESSA FILE, The
OEDIPUS NO HA

OFELAS
OFF LIMITS
OFFICER AND A GENTLEMAN,
  An
OGINSMA
OH GOD
OH GOD, YOU DEVIL
OH GOD II
OH WHAT A LOVELY WAR
OIRAN
OKLAHOMA CRUDE
OLD BOY FRIENDS
OLD DRACULA
OLIVER
OLIVER TWIST
OMAR MUKHTAR
OMEGA MAN, The
OMEN, The
ON A CLEAR DAY YOU CAN
  SEE FOREVER
ON GOLDEN POND
ON HER MAJESTY'S SECRET
  SERVICE
ON NE MEURT QUE 2 FOIS
ON THE DOUBLE
ONCE A THIEF
ONCE A THIEF (1965)
ONCE IS NOT ENOUGH
ONE, TWO, THREE
ONE AND ONLY, The
ONE FLEW OVER THE CUCK-
  OO'S NEST
ONE FOR THE DANCER
ONE IS A LONELY NUMBER
ONE MAGIC CHRISTMAS
ONE ON ONE
ONE SUMMER LOVE
ONKS VILJOO NÄKYNY
ONLY ONCE IN A LIFETIME
ONLY WHEN I LAUGH
ONYANKO CLUB
OPEN ADMISSIONS
OPERATION BOW WOW
OPERATION CROSSBOW
OPERATION THUNDERBOLT
OPTIMIST, The
ORANGE CURTAIN, The
ORCA - KILLER WHALE
ORDINARY PEOPLE
ORIENT EXPRESS
ORIENTAL EAGLE
ORPHANS
OTHELLO
OTHER, The
OTHER SIDE OF MIDNIGHT,
  The
OTHER SIDE OF MIDNIGHT II,
  The
OU EST LE PROBLEME?
OUGON NO PARTNER
OUR HOUSE
OUR MAN IN HAVANA
OUR TIME
OUR TIME IN HISTORY
OUR ZORA - NO SAMURAI

OURS, L
OUT OF BOUNDS
OUT OF TOWNERS, The
OUT ON A LIMB
OUTFIT, The
OUTLAND
OUTLAW JOSEY WALES, The
OUTRAGE, The
OUTRAGEOUS FORTUNE
OUTSIDERS, The
OVER THE TOP
OVERBOARD
OWL AND THE PUSSYCAT,
  The
OZORA NO SAMURAI

PACK, The
PACK OF LIES
PAINT YOUR WAGON
PAJAMA PARTY
PALE RIDER
PALTOQUET, Le
PANIC IN NEEDLE PARK
PAPA WAS A PREACHER
PAPER CHASE, The
PAPER HOUSE, The
PAPER MOON
PAPER TIGER
PAPILLON
PAPY FAIT DE LA RESISTANCE
PARADISE ALLEY
PARALLAX VIEW, The
PARANO
PARENTHOOD
PARIS OOH-LA-LA!
PARKING
PARTITA, La
PARTY, The
PASSAGE TO INDIA, A
PASSENGER, The
PASSION BEATRICE, La
PASSPORT TO OBLIVION
PAT GARRETT AND BILLY THE
  KID
PATCH OF BLUE, A
PATTI ROCKS
PATTON
PEACE IS HELL
PEANUT BUTTER SOLUTION,
  The
PEEPER
PEGGY SUE GOT MARRIED
PENELOPE
PENN & TELLER
PENN & TELLER GET KILLED
PENTE DOUCE, Le
PEPE
PERFECT
PERFECT COUPLE
PERFECT PROFILE
PERILS OF GWENDOLYN
PERIOD OF ADJUSTMENT
PERMISSION TO KILL
PERSIAN BLUE NO SHOZO
PET SEMATARY

PETE AND TILLIE
PETITE ALLUMEUSE, Le
PETITE AMIE, La
PETITE VOLEUSE, La
PETROS
PETULIA
PEYTON PLACE REVISITED
PHANTOM KILLER
PHANTOM OF THE MALL
PHANTOM OF THE PARADISE
PHANTOM OF THE RITZ
PHAR LAP
PHILOSOPHER KING, The
PHOTOGRAPHERS, The
PICK-UP ARTIST
PICK-UPS
PICNIC AT HANGING ROCK
PICTURE SHOW MAN, The
PIECE OF CAKE, A
PIECE OF THE ACTION
PIED PIPER, The
PIEGE/2 MN DE SOLEIL EN
  PLUS, Le
PIGEON THAT TOOK ROME,
  The
PILLOW TALK
PILOT, The
PINK CADILLAC
PINK FLOYD - THE WALL
PINK LADY
PINK PANTHER STRIKES
  AGAIN, The
PIPE DREAMS
PIPPIE LONGSTOCKING
PIRATE, The
PIRATE NI YOROSHIKU
PIRATES
PIRATES OF PENZANCE, The
PISTOLERO
PIT AND THE PENDULUM, The
PIZZA TRIANGLE, The
PLACE CALLED TODAY, A
PLANES, TRAINS &
  AUTOMOBILES
PLANET OF THE APES
PLAY DIRTY
PLAY IT AGAIN SAM
PLAY IT AS IT LAYS
PLAY MISTY FOR ME
PLAYERS
PLEASE DON'T EAT THE
  DAISIES
PLEASE NOT NOW!
PLEASE WHISPER MY NAME
PLENTY
POACHERS, The
POCKET MONEY
POCKETFUL OF MIRACLES
POINT BLANK
POINT DOULOUREUX, LE
POISON IVY
POLICE
POLICE ACADEMY
POLICE ACADEMY V
POLICE ACADEMY VI

262

POLICE NURSE
POLICE SQUAD/NAKED GUN
POLTERGEIST II
POLTERGEIST III
PONT DOUCE, Le
PONY EXPRESS RIDER
POOR LITTLE RICH GIRL
POPE JOAN
POPE OF GREENWICH VIL-
LAGE, The
POR ESO
PORK BUTCHER, The
PORRIDGE
PORTNOY'S COMPLAINT
PORTRAIT OF HELL
POSEIDON ADVENTURE, The
POSSE
POSSE POWER
POSSESSION OF JOEL
DELANEY
POULET AU VINAIGRE
POWER
PRAY FOR DEATH
PRAYER FOR THE DYING, A
PRAYING MANTIS
PREDATOR
PREDATORS
PREMATURE BURIAL, The
PREP SCHOOL
PRESIDENT'S ANALYST, The
PRESIDIO, The
PRESUMED INNOCENT
PRETTY IN PINK
PRETTY MAIDS ALL IN A
ROW
PRICE OF LIFE, The
PRICK UP YOUR EARS
PRIME CUT
PRINCE AND THE PAUPER,
The
PRINCE OF DARKNESS
PRINCE OF THE CITY
PRINCE PEACOCK
PRINCESS BRIDE, The
PRINCIPAL, The
PRISONER OF SECOND
AVENUE
PRISONER OF ZENDA
PRIVATE BENJAMIN
PRIVATE LIFE OF SHERLOCK
HOLMES, The
PRIZE, The
PRIZZI'S HONOR
PROFESSIONALS, The
PROJECT X
PROMISE, The
PROPHECY
PROTOCOL
PROVIDENCE
PSYCHO III
PT-109
PUBERTY BLUES
PUBLIC EYE
PUBLICATAIRE DUNLOPILLO
PULL-OVER ROUGE, LE

PUNCHLINE
PURPLE HEARTS
PURPLE RAIN
PURPLE ROSE OF CAIRO
PYRAMID

QUARTERMAIN
QUATERMASS
QUEENIE
QUEEN'S LOGIC
QUELQUES ARPENTS DE
NEIGE...
QUELQUES JOURS AVEC MOI
QUEST FOR FIRE
QUICK BEFORE IT MELTS
QUICK CHANGE
QUICKSILVER
QUILLER MEMORANDUM, The
QUILP
QUINTET

RABBIT RUN
RACE FOR THE YANKEE
ZEPHYR
RACE FROM OUTLAND
RACHEL'S MAN
RAD
RADFAU DE LA MEDUSE, Le
RADIO DAYS
RAFFERTY AND THE GOLD
DUST TWINS
RAGE
RAGE OF HONOR
RAGE TO LIVE, A
RAGGEDY ANN AND ANDY
RAGING BULL
RAGMAN'S DAUGHTER, The
RAIDERS OF THE LOST ARK
RAIN MAN
RAINTREE COUNTY
RAISE THE TITANIC
RAISING ARIZONA
RAISING DAISY ROTHCHILD
RAMBO: FIRST BLOOD II
RAMPAGE
RAN
RANCHO DELUXE
RARE BREED, The
RAT BOY
RAUL WALLENBERG
RAVAGES, The
RAVEN, The
RAY'S MALE HETEROSEXUAL
DANCE HALL
RAZORBACK
RE DEI SETTE MARI, IL
RED GAMES
REAL GENIUS
REAL LIFE
REAL MEN
REBEL
REBEL LOVE
RECKLESS ENDEARMENT
RED, WHITE & BLACK
RED DAWN

RED HEADED STRANGER
RED HEAT
RED ROSE
REDNECK
REFLECTION OF FEAR
REFLECTIONS IN A GOLDEN
EYE
REGATTAS OF SAN
FRANCISCO
REGLEMENT DE COMPTES
REIN
REINCARNATION OF PETER
PROUD, The
REIVERS, The
REMANDO AL VIENTO
REMEMBER MY NAME
REMEMBER THAT POKER
PLAYING MONKEY
REMO WILLIAMS
RENEGADE
RENGOU KANTAI
RENT-A-COP
REPEAT "WHEN 8..."
REPORT TO THE
COMMISSIONER
REPUBLIC OF UZICE
RESIDENCE SURVEILLEE
RESTING PLACE
RETENEZ MOI OU JE FAIS
MALHEUR
RETURN FROM THE ASHES
RETURN OF A MAN CALLED
HORSE
RETURN OF CAPTAIN
INVINCIBLE
RETURN OF SHERLOCK
HOLMES, The
RETURN OF SNOWY RIVER
RETURN OF THE
BOOMERANG
RETURN OF THE LIVING DEAD
II
RETURN OF THE PINK
PANTHER
RETURN OF THE SEVEN, The
RETURN OF THE SOLDIER,
The
RETURN TO BOGGY CREEK
RETURN TO MACON COUNTY
REUNION
REVENGE
REVENGE OF THE NERDS
REVENGE OF THE NERDS II
REVENGE OF THE PINK
PANTHER
REVENGERS, The
RHINESTONE
RHINOCEROS
RIATA
RICCO
RICH AND FAMOUS
RICHARD'S THINGS
RIDDLE OF THE SANDS,
The
RIDE THE HIGH COUNTRY

263

RIGHT STUFF, The
RIMINI RIMINI UN ANNO
   DOPO
RING, The
RING OF FIRE
RIO QUE NOS LLEVA, El
RIPOUX, Les
RITZ, The
RIVER, The
RIVER NIGER
RIVER RATS
RIVERS OF STEEL
RIVIERA
ROAD HOUSE
ROAD TO SALINA, The
ROAD WARRIORS
ROADIE
ROAR
ROBBIE ZENITH
ROBIN AND MARION
ROBIN AND THE SEVEN
   HOODS
ROBINSON CRUSOE
ROCK YO SHIZUKANI NAGERO
ROCKET TO THE MOON
ROCKULA
ROCKY II
ROCKY III
ROCKY IV
ROKUMEIKAN
ROLLERBALL
ROLLERCOASTER
ROLLOVER
ROMAN HOLIDAY (1988)
ROMANCING THE STONE
ROMANTIC ENGLISH WOMAN,
   A
ROMUALD ET JULIETTE
RONDE DE NUIT
ROOFTOPS
ROOSTER COGBURN
ROSE, The
ROSEBUD
ROSELAND
ROTTEN TO THE CORE
ROUGH RIDERS
ROUND MIDNIGHT
ROUNDERS, The
ROXANNE
ROYAL FLASH
RUBA AL PROSSIMO TUO
RUBICON
RUMBA, La
RUNAWAY
RUNNER STUMBLES, The
RUNNERS
RUNNING MAN, The (1963)
RUNNING MAN, The (1987)
RUNNING ON EMPTY
RUNNING SCARED
RUSA, La
RUSSIANS ARE COMING, THE
   RUSSIANS ARE COMING,
   The
RUSSICUM

RUSTLER'S RAPSODY
RUTHLESS PEOPLE
RYAN'S DAUGHTER

S.O.B.
S.O.S. TITANIC
SABLE
SABOR DE LA VENGANZA, El
SAFARI 3000
SAHARA SECRET, The
SAIGON
SAIGONO RAKUEN
SAILOR, The
SAILOR WHO FELL FROM
   GRACE WITH THE SEA, The
SAKHAROV
SALOME
SALUTE OF THE JUGGLER
SAME TIME NEXT YEAR
SAMMIE AND ROSIE GET LAID
SANCTUARY
SAND PEBBLES, The
SANDPIPER, The
SANDPIT GENERALS, The
SANDS OF THE KALAHARI,
   The
SANG DES AUTRES, Le
SANGRE Y ARENA
SANTA FE "1836"
SANTOU KOUKOUSEI
SARABA HAKOBUNE
SARAH
SATAN BUG, The
SATISFACTION
SATURDAY NIGHT FEVER
SATURN 3
SATYRICON
SAVAGE IS LOOSE, The
SAVAGE SWARM
SAVE THE TIGER
SAXO
SAY ANYTHING
SAYONARA JUPITER
SCALP HUNTERS
SCANDAL
SCANDALOUS
SCANDALOUS ADVENTURES
   OF BARAIKAN, The
SEVEN YEAR STORM
SCANDALOUS JOHN
SCARECROW
SCARFACE
SCARLET PIMPERNELL
SCHLOCK
SCOUTS TOUJOURS
SCROOGE
SCROOGED
SEA OF LOVE
SEA WOLVES
SEAL KIDS, The
SECOND SIGHT
SECOND VICTORY, The
SECRET DU LAC D'ARGENT,
   Le
SECRET FILES

SECRET GARDEN, The
SECRET INVASION, The
SECRET LIFE OF PLANTS, The
SECRET OF MY SUCCESS
   (1965), The
SECRET OF MY SUCCESS
   (1987), The
SECRET OF SANTA VITTORIA,
   The
SECRET OF THE PLANET
   EARTH
SEE NO EVIL
SEE YOU IN THE MORNING
SEGOVIA
SEINEN NO KI
SEISHOKU NO ISHIBUMI
SEISYOKU
SEMAIN DE VACANCES, UNE
SEMI-TOUGH
SENDER, The
SENIOR PROM
SENSOU O SHIRANAI
   KODOMOTACHI
SENTINEL, The
SEPARATE PEACE, A
SEPTEMBER
SEPTIEME CIBLE, La
SERGEANT'S THREE
SERGEANT DEADHEAD
SERGEANT PEPPER'S LONELY
   HEARTS CLUB BAND
SERIAL
SERIE NOIRE
SERPENT, The
SERPENT A PLUMES, Le
SERPICO
SETOUCHI SHONEN
   YAKYUDAN
SETSATOMI HAKKENDEN
SEVEN-UPS, The
SEVEN GOLD VAMPIRES
SEVEN MEN AT DAYBREAK
SEVEN MINUTES
SEVEN NIGHTS IN JAPAN
SEVEN PERCENT SOLUTION,
   The
SEVEN WOMEN
SEVEN YEAR STORM, The
SEX, LIES AND VIDEOTAPE
SEXWORLD
SHADOW HORSE
SHADOW MOUNTAIN
SHADOW ON THE STORM
SHAFT
SHAFT IN AFRICA
SHAFTS BIG SCORE
SHAKE & BAKER
SHALIMAR
SHAMPOO
SHAMUS
SHARKEY'S MACHINE
SHE DEVIL
SHE'S HAVING A BABY
SHE CAME TO THE VALLEY
SHEENA QUEEN OF THE
   JUNGLE

264

SHELLY
SHERLOCK AND ME
SHERLOCK HOLMES'
  SMARTER BROTHER
SHIBAMATABORI AIOKOMETE
SHINSHI DOMEI
SHIP OF FOOLS
SHIRETOKO BOJO
SHIRLEY VALENTINE
SHOCK TREATMENT
SHOCKWAVE
SHOELESS JOE
SHOES OF THE FISHERMAN
SHOGUN ASSASSIN
SHONBEN RIDER
SHOOT THE MOON
SHOOTER
SHOOTING PARTY, The
SHOOTING STARS
SHOOTIST, The
SHORT CIRCUIT
SHORT CIRCUIT II
SHORT EYES
SHOSTAKOVICH
SHOT IN THE DARK, A
SHOUT, The
SHOUT AT THE DEVIL
SICILIAN CLAN, The
SIDDHATHA
SIDE ROAD
SIDEWINDER 1
SIGN POST TO MURDER
SIGNE CHARLOTTE
SILENCE OF THE LAMBS, The
SILENT MOVIE
SILENT NIGHT
SILENT RAGE
SILKWOOD
SILVER DREAM RACER
SILVER STREAK, The
SILVERADO
SINBAD
SINFUL DAVY
SINGING NUN
SINGLE ROOM
SINS
SIR GAWAIN AND THE GREEN
  KNIGHT
SISTER SISTER
SIX AGAINST THE ROCK
SIX PACK
SIX WEEKS
SIXTEEN CANDLES
SKI FANTASY
SKI PARTY
SKI RAIDERS, The
SKIDOO
SKIN DEEP
SKIN GAME
SKIN HEADS
SKIPPING
SKULLDUGGERY
SKYJACK
SLAEGTEM
SLAMS

SLAPSHOT
SLAUGHTERHOUSE
SLAVES OF NEW YORK
SLEEP WELL MY LOVE
SLEEPER
SLEUTH
SLIM DUSTY MOVIE, The
SLIPPER AND THE ROSE, The
SLIPSTREAM
SLITHER
SLUGGER'S WIFE
SMILE
SMOKEY & THE BANDIT III
SMOKEY AND THE BANDIT
SNAKE HEAD
SNAPSHOT
SNEAKER BLUES
SÖNDROTTINGEN
SNOW JOB
SNOW QUEEN, The
SO I'LL GO QUIETLY
SO WHAT
SOKOLOVE
SOL MADRID
SOLAR BABIES
SOLDIER, The
SOLDIER'S STORY
SOLDIER BLUE
SOLDIER BOY
SOME CAME RUNNING
SOME KIND OF WONDERFUL
SOMEBODY KILLED HER
  HUSBAND
SOMEONE TO WATCH OVER
  ME
SOMETHING IN COMMON
SOMETHING LIKE THE TRUTH
SOMETIMES A GREAT
  NOTION
SOMEWHERE IN TIME
SONG OF NORWAY
SONG WITHOUT END
SONGE DES AUTRES, Le
SONNET FOR THE HUNTER, A
SONNY BOY
SONS OF KATIE ELDER
SONS OF SASSOUN
SOPHIE'S CHOICE
SORCERER
SOREKARA
SOROBAN JYUKU
SOS
SOSHUN MONOGATARI
SOTSUGYOU RYOKOU
SOTTOZERO
SOUL MAN
SOUL OF NIGGER, The
SOUNDER
SOUS, Le
SOUTH PACIFIC
SOUTHERN COMFORT
SOUVENIRS SOUVENIRS
SOYLENT GREEN
SPACE CAMP
SPACEBALLS

SPARKLE
SPECIAL DELIVERY
SPECIAL POLICE
SPECIALISTES, Les
SPEED LIMIT 65
SPEEDWAY
SPENCERS MOUNTAIN
SPHINX
SPIDER MAN
SPINOUT
SPIRALE
SPIRALS
SPLASH
SPLATTER
SPLIT
SPOOK THAT SAT BY THE
  DOOR, The
SPORTING PROPOSITION
SPY WHO LOVED ME, The
SQUEEZE, The
ST. ELMO'S FIRE
ST. VALENTINE'S DAY
  MASSACRE
STAIRCASE
STAKEOUT
STALKING MOON
STALLIONS
STAND BY ME
STANLEY & IRIS
STAR IS BORN (1976), A
STAR TREK - THE MOTION
  PICTURE
STAR TREK II
STAR TREK III
STAR TREK IV
STAR TREK V
STAR WARS
STARDUST
STARDUST MEMORIES
STARMAN
STARTING OVER
STATE OF GRACE
STATION
STAY AWAY JOE
STAY HUNGRY
STAYING ALIVE
STEALING HOME
STEEL MAGNOLIAS
STEELYARD BLUES
STELLA DALLAS
STEPFORD WIVES, The
STEWARDESS SCHOOL
STICK
STICKY FINGERS
STILL OF THE NIGHT
STING, The
STINGRAY
STIR CRAZY
STORMIN HOME
STORY OF CINDERELLA, The
STORY OF JOSEPH & HIS
  BRETHREN
STRAIGHT FROM THE HEART
STRANDED
STRANGE BEHAVIOR
STRANGE INVADERS (1983)

STRANGE SHADOWS IN AN
  EMPTY ROOM
STRANGER AND THE GUN-
  FIGHTER, The
STRANGERS WHEN WE MET
STREET LAW
STREET WISE
STREETS OF FIRE
STREETS OF GOLD
STREETS OF MALICE
STREGATI
STRETCH HUNTER
STRIKING BACK
STRIPPER, The
STRONG MEDICINE
SUBMARINE COMMAND
SUBTERRANEANS, The
SUCH GOOD FRIENDS
SUDDEN IMPACT
SUEÑO DEL MONO LOCO, El
SUGARLAND EXPRESS
SUGATA SANSHIROU
SUMMER AND SMOKE
SUMMER FIRES
SUMMER OF '42
SUMMER RENTAL
SUMMER SCHOOL
SUMMER WISHES, WINTER
  DREAMS
SUMMERFIELD
SUMMERTIME YANKS
SUNDAY BLOODY SUNDAY
SUNSET
SUNSHINE BOYS, The
SUPER COPS, The
SUPERFLY
SUPERFLY T.N.T.
SUPERGIRL
SUPERMAN I
SUPERMAN II
SUPERMAN III
SUPPORT YOUR LOCAL
  GUNFIGHTER
SUR LA ROUTE DE SALINA
SURVIVORS, The
SUTJESKA - THE FIFTH
  OFFENSIVE
SWAMP RATS
SWARM, The
SWASHBUCKLER
SWEET BIRD OF YOUTH
SWEET CHARITY
SWEET CREEK COUNTY WAR,
  The
SWEET DREAMS
SWEET HOME
SWEET RIDE
SWEET SUZY
SWEETHEART'S DANCE
SWEETHEARTS
SWEETWATER
SWIMMING TO CAMBODIA
SWING SHIFT
SWISS CONSPIRACY
SWISS FAMILY ROBINSON

SWITCHING CHANNELS
SWORD AND THE SORCERCER
SWORD OF LANCELOT
SWORD OF VENGENCE
SWORDSMAN OF SIENA

T.R. BASKIN
TABIJI
TABITO ONNATO TORAJIROU
TABLE FOR FIRE
TAKE HER, SHE'S MINE
TAKE THE MONEY AND RUN
TAKETORI MONOGATARI
TAKING OF PELHAM 1-2-3
TALES OF NUNUNDAGE
TALES OF TERROR
TALK RADIO
TAM LIN
TAMARIND SEED
TAMING OF THE SHREW
  (1967)
TAN TAN TANUKI
TANK, The
TANPOPO
TANT PIS SI JE MEURS
TANTEI MONOGATARI
TAP
TARAS BULBA
TARGET
TAROTS
TARZAN AND THE GREAT
  RIVER
TARZAN AND THE JUNGLE
  BOY
TARZAN AND THE VALLEY OF
  GOLD
TARZAN GOES TO INDIA
TARZAN IN BRAZIL
TASTE FOR FEAR, A
TASTE OF SAVAGE
TAXI BOY
TAXI DRIVER
TELEFON
TELEPHONE, The
TELL THEM WILLIE BOY IS
  HERE
TEMPEST
TEN-SECOND JAIL-BREAK, The
TEN GIRLS AGO
TENANT, The
TENDER, The
TENDER MERCIES
TENTH MAN, The
TENUE DE SOIREE
TEQUILA SUNRISE
TERMINAL MAN
TERMS OF ENDEARMENT
TERRACITTA WARRIORS &
  HEROES FROM THE MAU-
  SOLEUM OF QIN SHI
  HUANG
TERRIBLE BEAUTY, A
TERROR CIRCUS
TESS

TESTA DURA
TESTIMONY
TEXAS PICTURE, The
TEXASVILLE
THANK YOU SATAN
THAT'S COUNTRY
THAT'S ENTERTAINMENT
  TOO!
THAT'S LIFE
THAT'S THE WAY OF THE
  WORLD
THAT TOUCH OF MINK
THE YAKUZA
THEO MATI
THERE GOES THE BRIDE
THERE WAS A CROOKED MAN
THEY CAME TO CORDURA
THEY LIVE
THEY ONLY KILL THEIR
  MASTERS
THEY SHOOT HORSES, DON'T
  THEY
THEY STILL CALL ME BRUCE
THIEF OF BAGHDAD
THIEF OF HEARTS
THIEF WHO CAME TO DINNER,
  The
THIEVES LIKE US
THING, The (1982)
THINGS ARE TOUGH ALL
  OVER
THINGS CHANGE
THIRD DATE, The
THIRD DAY, The
THIRST
THIRTY-SIX HOURS
THIS EARTH IS MINE
THIS PROPERTY IS
  CONDEMNED
THOMAS CROWN AFFAIR, The
THOSE DARING YOUNG MEN
THOSE DEAR DEPARTED
THOSE FANTASTIC FLYING
  FOOLS
THOSE LIPS, THOSE EYES
THREE AMIGOS
THREE BITES OF THE APPLE
THREE DAYS OF THE CONDOR
THREE FOR ALL
THREE MEN & A BABY
THREE MUSKETEERS, The
THREE O'CLOCK HIGH
THREE SISTERS
THREE WOMEN
THROW MOMMA FROM THE
  TRAIN
THROWBACK
THUNDER ALLEY
THUNDER OF DRUMS, A
THUNDERBALL
THUNDERBOLT AND
  LIGHTFOOT
TI PRESENTO UN'AMICA
TICK, TICK, TICK
TICKLE ME
TICKLISH AFFAIR, A

267

X-15
Y A BON LES BLANCS
YABANJINNO YONI
YARINO GONZA
YASHA
YEAR OF LIVING DANGER-
 OUSLY, The
YEAR OF THE DRAGON
YELLOW CANARY, The
YELLOW DOG
YELLOW ROLLS ROYCE
YELLOWBEARD
YELLOWHAIR
YES. GIORGIO
YOAKENO RUNNER
YOG - MONSTER FROM
 SPACE
YOGIRINI MUSEBU TORAJIRO

YOGORETA EIYU
YOU ARE MY DESTINY
YOU CAN'T WIN THEM ALL
YOU ONLY LIVE TWICE
YOU TALKING TO ME
YOUNG EINSTEIN
YOUNG FRANKENSTEIN
YOUNG GUNS
YOUNG REBEL
YOUNG RUNAWAYS
YOUNG SHERLOCK HOLMES
YOUNG WINSTON
YOUR CHEATIN HEART
YOUR THREE MINUTES ARE
 UP
YUKI NO DANSHOU

YUSHUN
ZABRISKIE POINT
ZACK
ZANDYS BRIDE
ZANSHOU
ZARDOZ
ZATOICHI
ZATOICHI AT LARGE
ZEGEN
ZELIG
ZEPPLIN
ZEROSEN MOYU
ZIG ZAG
ZOKU NINGENKUMEI
ZOOM
ZOOT SUIT
ZULU DAWN

Index

270

271